THE STUDY OF VARIABLE STARS USING SMALL TELESCOPES

Edited by

John R. Percy

Erindale College and
David Dunlap Observatory
Department of Astronomy
University of Toronto

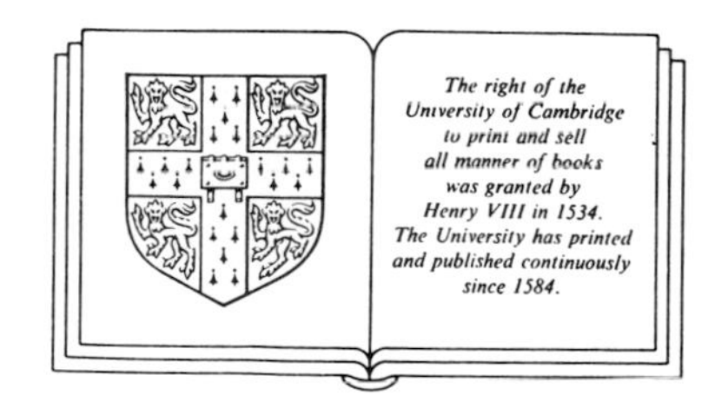

CAMBRIDGE UNIVERSITY PRESS

Cambridge
London New York New Rochelle
Melbourne Sydney

Published by the Press Syndicate of the University of Cambridge
The Pitt Building, Trumpington Street, Cambridge CB2 1RP
32 East 57th Street, New York, NY 10022, USA
10 Stamford Road, Oakleigh, Melbourne 3166, Australia

First published 1986

British Library cataloguing in publication data

The study of variable stars using small telescopes.

1. Stars, Variable
I. Percy, John R.
523.8′44 QB835

Library of Congress cataloguing in publication data available

ISBN 0 521 33300 8

TABLE OF CONTENTS

PREFACE

This symposium on "The Study of Variable Stars using Small Telescopes" was held at the University of Toronto from July 11 to 14, 1985, as part of the fiftieth anniversary celebrations of the university's David Dunlap Observatory. The symposium was intended for a broad audience, including professional and amateur research astronomers, and astronomy teachers at universities with small telescopes. These proceedings are intended for the same audience. The symposium was built around a series of invited reviews by astronomers who are not only experts in their field, but are also sympathetic to the problems and potential of small-telescope astronomy. All of these reviews are included in these proceedings, along with several short "case studies" which effectively complement them. Unfortunately, these proceedings do not record the interesting and useful discussions which took place after the lectures, at coffee breaks and at other social events during the symposium. However, the style of many of the papers captures something of the friendly informality of the meeting.

The symposium was co-sponsored by the American Association of Variable Star Observers (AAVSO), International Amateur-Professional Photoelectric Photometry (IAPPP) and the Royal Astronomical Society of Canada (RASC). The work of the AAVSO and IAPPP is described in detail within this book. The RASC does not carry out variable star work on an institutional basis, but encourages its members to join the AAVSO or IAPPP, and several of them have made notable contributions to variable star research through these organizations. The history of the RASC goes back more than a century. It now has about 3000 members, both amateur and professional. Most are attached to one of the 20 centres across Canada, but there are several hundred unattached members around the world.

The David Dunlap Observatory has been involved in variable star research throughout its fifty-year history. The study of variable stars has also been incorporated into the teaching program at the university, in the form of course and thesis projects for undergraduate and graduate students. We think that the study of variable stars provides an ideal introduction to the concepts of measurement, analysis and interpretation of scientific data. Instructors at other universities may wish to develop similar projects for their students — projects which involve real observations which contribute to the advancement of scientific knowledge.

The study of variable stars is an area in which the amateur can also make a contribution to science. This possibility is discussed in several papers in this book, perhaps most eloquently in Chris Sterken's. Many of the staff of the David Dunlap Observatory have recognized this contribution through their strong support of the AAVSO, IAPPP and RASC. For the amateur, there is a special satisfaction in discovering a nova, being a co-author of a scientific paper, recognizing a contributed point on a published light curve, or simply knowing that one's hobby has contributed to scientific knowledge. We hope that this book provides insight and inspiration to those who are already observing variable stars, and encouragement to others who would like to enter this field. There is no shortage of work to be done; this will be evident to anyone who reads this book.

The emphasis in this book is on science and scientific programs. For information on other aspects of variable stars, the reader should consult the books listed in the various reviews, particularly those by Doug Hall and Ed Guinan.

ACKNOWLEDGEMENTS

The scientific organizing committee for the symposium consisted of Emilia Belserene, Christine Clement, Russ Genet, Doug Hall, Janet Mattei and Ian McGregor, with the undersigned as chairman. The local organizing committee included Christine Clement, Alex Fullerton, Ian McGregor, Joan Tryggve and myself. Many other members of the Department of Astronomy assisted with the local arrangements in various ways, and we are very grateful for their help — especially since this was the fourth meeting in the department in less than two months!

The organizers of this symposium wish to thank the University of Toronto for playing host, and particularly Donald Fernie for providing a grant from the Carl Reinhardt Visiting Lecturer Fund to partially cover the travel expenses of the invited review speakers. A grant from the Natural Sciences and Engineering Research Council of Canada helped to cover the cost of the preparation of these proceedings.

I wish to thank Ann Rusk for typing and formatting the text of these proceedings with speed, care and judgement, and Simon Mitton and the Cambridge University Press for agreeing to publish it. May it repay the efforts of all the contributors and helpers by encouraging more and better observations of variable stars, and providing inspiration, enlightenment and enjoyment to its readers.

John R. Percy
May, 1986

THE STUDY OF VARIABLE STARS AT THE DAVID DUNLAP OBSERVATORY

John R. Percy & J. Donald Fernie
David Dunlap Observatory
Box 360
Richmond Hill, Ontario
Canada L4C 4Y6

The David Dunlap Observatory (DDO) is a centre for research in astronomy, operated by the Department of Astronomy, University of Toronto. It is located about 25 km north of central Toronto, in the town of Richmond Hill, Ontario. This town, once separated from the city by farmland, is now within the city's expanding fringe, and "light pollution" has been a problem for some time. Nevertheless, the observatory enjoys about 1200 hours of "spectroscopic" observing conditions each year. About half these hours are also suitable for photometry. By judicious choice of scientific programs, and some technological ingenuity, the observatory continues to do worthwhile research. It also contributes to the training of graduate students in astronomy, and attracts thousands of visitors to its public and school tours each year.

Observatories such as the DDO have a special role to play in modern astronomical research. Although it is true that much of present-day observational astronomy is done at sites which are geographically remote but climatologically superb, there is heavy demand for the use of telescopes at these sites: everyone wants to use them. As a result, an astronomer may typically receive only a few nights of observing time each year. This has led to a tendency for astronomers to work on projects which can be completed in a few concentrated nights of observation. Projects that require the patient accumulation of data over long periods of time cannot be done there, scientifically rewarding though they might be. Local observatories such as the David Dunlap Observatory, even at mediocre sites, have several distinct advantages in terms of the scientific projects which they can undertake, and they can therefore help to restore the balance.

1. They can undertake large-scale surveys, such as spectral classification and radial velocities, which among other things identify stars worthy of individual, detailed study.

2. They can obtain intensive, complete phase coverage of regular, short-period phenomena (binary orbits, variable stars) because, over a long time interval, all phases can eventually be filled in.

3. They can identify and study cycle-to-cycle changes in irregular, short-period phenomena: RS CVn binary orbits, for instance.

4. They can study slow phenomena: long-period binaries and variable stars such as novae and Miras.

5. With flexibility of scheduling, they can study unexpected events such as novae, and can participate with other observatories in co-ordinated, multi-technique "campaigns" on specific stars or groups of stars; they can also follow up immediately on any discovery made in the course of an ongoing observing program.

In the rest of this article, we will describe the research work on variable stars at the DDO, and show how it exploits the special advantages of a local observatory.

The History of the David Dunlap Observatory

This meeting celebrates the fiftieth anniversary of the DDO. Although the observatory was opened in 1935, its origins lie many decades earlier. It was the dream and life's ambition of one man — Clarence Augustus Chant — and its opening, coincident with his seventieth birthday, marked the culmination of his life's work.

Chant was born in 1865, not far from where the observatory now stands. He graduated from the University of Toronto in 1890 and, shortly afterward, he joined the university as Lecturer in Physics. Here he spent the remainder of his long career, until he passed away in his ninety-second year. By 1904, he had convinced the university to establish a small Department of Astronomy (with himself as the only member), and to introduce astronomy as an option in the mathematics and physics curriculum. As early as 1908, he was writing of the importance to the university — and indeed to the city — of acquiring a major astronomical observatory.

His task proved to be a difficult one, fraught with one disappointment after another. In 1926, however, his prayers were answered. Jessie Donalda Dunlap agreed to establish a major astronomical observatory as a memorial to her late husband, a lawyer with a keen interest in astronomy. A site was chosen — in a casual fashion which is quite in contrast with today's procedures — and on May 31, 1935, the observatory was officially opened.

From the start, it had been decided that the great 1.9 m (74") telescope would be devoted mainly to spectroscopic studies of stars. Such studies provide astronomers with much information about the chemical composition of the stars, as well as the physical conditions such as temperature and pressure which prevail in their atmospheres. In the course of large-scale spectroscopic studies, many abnormal stars were discovered. Two types — the peculiar A stars and the Be stars — are variable in spectrum and brightness, and many such stars have been followed up with detailed study. Spectroscopic studies also provide information on stellar motions, through the Doppler effect. A by-product of these studies has been the discovery and analysis of dozens of spectroscopic binary stars, especially by John F. Heard and Ruth J. Northcott. Many of these systems have proven to be eclipsing binaries as well. Late in 1948, experiments in photoelectric photometry were begun at the observatory, using a 0.5 m reflector built by Reynold K. Young in 1929. This telescope has since been used for studies of numerous variables of all kinds.

A variety of important changes occurred at the observatory, beginning about 1960. The size and diversity of the staff increased, in response to the growing enrolment at the university. Some work in theoretical astrophysics and radio astronomy had begun as early as the 1940's; this expanded greatly in the 1960's. Donald A. MacRae, the fifth director, played a significant role in the development of radio astronomy at the university. The university's first radio telescopes were in the back yard of the DDO, and one of their first projects was to measure the declining radio flux from a "variable star" — the remnant of the supernova Cas A. MacRae was also responsible for the first astronomical use of the university's computers — for determining orbits of spectroscopic binaries. The computers were later used for theoretical studies, particularly some on stellar pulsation by Pierre Demarque, John R. Percy and their students. The work of the small but very active radio astronomy group has included some important studies of "radio variable stars" (novae, supernovae, Be stars and mass-transfer binaries including symbiotic stars) by Ernest R. Seaquist and his students. There was interest in infrared astronomy as early as the 1960's, but this developed more fully in the 1970's with the arrival of Robert A. McLaren. A monumental study of Cepheids at infrared wavelengths began under the direction of him and Barry F. Madore, and is described in some detail in the article by Welch elsewhere in this volume.

Another arrival at the DDO in the 1960's was the 0.6 m telescope. This rugged and versatile instrument has been used for a variety of spectroscopic and photometric studies, many of them described below. It has also been used for instrument-testing for a more remote 0.6 m telescope: in 1970, the observatory decided to establish a southern field station at the edge of the Atacama Desert in Chile. Here, high in the foothills of the Andes, are to be found some of the clearest and darkest skies anywhere on Earth. Much of the observational work of the observatory is carried out at this and other remote sites — the Canada-France-Hawaii observatory, Kitt Peak and Cerro Tololo observatories among others. The DDO remains the "home base". There, astronomers have access to analytical facilities such as the PDS microdensitometer, and sophisticated image-processing devices. As of this date (1986), the observatory hopes to expand its observational facilities in Chile as well as its analytical facilities in Richmond Hill.

The observational facilities at the "home base" are certainly not being neglected. New detectors are being constructed for the 1.9 m telescope, to provide for greater sensitivity. They also provide for subtracting the sky contribution from the signal being recorded. A similar principle has been applied to the observatory's photometric studies. The 0.5 and 0.6 m telescopes work in tandem, under computer control, one measuring a variable star and the sky around it, and the other measuring a comparison star and sky. This procedure removes the effects of both sky brightness and haze, and makes it possible to do photometric studies on almost twice as many nights as before.

Variable Stars in Globular Clusters

Beginning in 1935, a small but important fraction of the observing time on the 1.9 m telescope was devoted to the photographic observation of variable stars in globular clusters. This work was carried out by Helen Sawyer Hogg, who had begun her studies of these stars at Harvard College Observatory under the supervision of Harlow Shapley, and had continued it at the Dominion Astrophysical Observatory in Victoria. These studies continued for five decades, though after 1970, the observational part of the program was transferred to Las Campanas. Most of the work on this program is now done by Christine Clement, who carried out her doctoral research under Helen Hogg's supervision, and who has been a staff member of the DDO since 1969.

The study of variable stars in globular clusters is important for many reasons. The globular clusters are the oldest groupings of stars in our galaxy, having ages of 10 billion years or more. Most of the variables in them are RR Lyrae stars, which undergo radial pulsation in either the fundamental or first overtone mode. The intrinsic luminosity of the RR Lyrae stars can be determined in a variety of ways. Then, if they can be identified and measured in globular clusters, the precise distance of the clusters can be determined. More recently, Arthur N. Cox and others have found that some RR Lyrae stars pulsate in the fundamental and first overtone modes simultaneously, and that the ratio of the two periods can be used to infer the mass and chemical composition of these stars (see article by Clement & Nemec elsewhere in this volume). Finally, the periods of pulsation are subject to small changes which may be due to the evolutionary changes in the radii of the stars. These changes are cumulative — like the errors in a clock — and if they are observed over a sufficiently long period of time, they can provide us with the only direct evidence of the normal, slow evolution of the stars.

In 1972, an International Astronomical Union Colloquium on Variable Stars in Globular Clusters was held at the University of Toronto in honour of Helen Sawyer Hogg's retirement from the university, and in recognition of the importance of her lifelong work (Fernie 1973).

Stars with Spots and Rings

The spectroscopic study of binary stars has long been a specialty of the DDO, and it is fitting that the observatory's most newsworthy discovery was in this area. Bolton (1972) presented evidence for the existence of a black hole accompanying the star HDE 226868, identified with the X-ray source Cygnus X-1. This was the first known example of a probable black hole in space. Numerous other binary stars have been studied, both to determine their "vital statistics" and to investigate mass transfer and circumstellar matter within the systems. The ability of a local observatory to obtain complete phase coverage of a binary orbit is essential for studies of these kinds. An interesting product of this work on spectroscopic binaries was a paper by Tanner (1948) on the occurrence of spurious periods in binary and variable stars.

The earliest photometric studies at the DDO were carried out with the 0.5 m telescope, and dealt with eclipsing binaries which were also being studied spectroscopically. Usually, the eclipse light curves were quite normal, and routinely provided information about the luminosities, temperatures, radii and masses of the stars. A few of these binaries were more problematic: there were pronounced variations which were not due to eclipses, and these variations changed from cycle to cycle. Such stars are now called RS CVn stars, and their variations are known to be due to starspots (Hall 1976), but they were studied at the DDO long before they obtained their present name and notoriety. Herbst (1973) carried out systematic observations of a few of them as an M.Sc. thesis project. One or two of these stars had first been observed at the DDO almost two decades earlier. Important spectroscopic studies of RS CVn stars have also been carried out, most recently by Dorothy A. Fraquelli.

Stars with Variable Spectra

There are many stars whose most pronounced variability is spectroscopic, and these have been of particular interest to some astronomers at the DDO. A few novae have been studied, and several peculiar A stars (also known as "spectrum variables"). These are stars with strong magnetic fields, which are inclined at an oblique angle to the axis of rotation of the star. One of the effects of the magnetic field is to modify the distribution of the chemical elements over the surface of the star, so that the appearance (spectrum, brightness, colour) of the star depends on the orientation of the axis of rotation, the magnetic field and the observer. As a result, the spectrum, brightness and colour of the star change as the star rotates. The helium-variable B stars, which are close relatives of the peculiar A stars, have been intensively observed in recent years.

Other work on stars with variable spectra has concerned the Be (emission) stars (see article by Percy elsewhere in this volume). These stars are numerous, interesting and enigmatic. The cause of the emission, and its variability, has not yet been determined. In a study ideally matched to the facilities of the DDO, John F. Heard systematically monitored the spectroscopic variability of 60 Be stars over 24 years (Copeland & Heard 1963).

Current interest in Be stars at the DDO centres on their short-period variability, which is believed to be due to pulsation. Changes in the amplitude of the pulsation may be the cause of the long-term spectroscopic variability — a possibility which was first proposed by Bolton (1982) on the basis of a spectroscopic study of λ Eri carried out at the DDO.

Photometric Studies of Pulsating Variables

With the coming of Donald Fernie in 1960, and the coming of the 0.6 m telescope shortly after, photometric research at the DDO expanded greatly. Several large-scale studies of variable stars were begun, many of them as graduate thesis projects. For reasons mentioned earlier, the climate and facilities at the DDO are admirably matched to projects on long-period and irregular variables, which typically take a year or two to complete — the typical time scale of a thesis project. Furthermore, the availability of observing time is not subject to the whims of a time allocation committee; large blocks of time are available for survey projects.

One such project uncovered "the case of the reluctant Cepheid". In the course of a survey of Population II Cepheids, Serge Demers discovered that one of them — RU Cam — had ceased to vary (Demers & Fernie 1966). The explanation for this unique behaviour is still not clear, but it seems to reflect the rather transient evolutionary state of these stars. Among the many other unusual Cepheids studied at the DDO over the years, the best-known is probably the North Star — Polaris. The amplitude of this 4-day Cepheid has gradually decreased over the last few decades, and is now only about 0.03 magnitude. These results demonstrate the value of monitoring variable stars on a routine, long-term basis.

Nancy R. Evans (1970) investigated the magnitudes and colours of Mira variables at maximum light. Subsequently, Thomas G. Barnes (1973) carried out a massive survey of more than 175 Mira variables which exploited the advantages — brightness and reduced amplitude — of observing these stars in the near-infrared VRI bands (see the article by Wing elsewhere in this volume). David L. DuPuy (1973) surveyed the RV Tau stars. Studies of these Cepheid-like stars are hampered by their long cycle times (up to 150 days) and by their cycle-to-cycle variations. DuPuy also contributed to an important study of R CrB stars (Fernie, Sherwood & DuPuy 1972), which not only provided information about the characteristic fadings of these stars, but also provided the first evidence for pulsation in some of these stars — notably R CrB itself.

Classical (Population I) Cepheids have been the subject of numerous papers, dealing especially with their distances and their use as a probe of galactic structure. Nancy R. Evans (1976a, b) used contemporaneous spectroscopic and photometric observations to determine the radii of many of these stars, by means of the Baade-Wesselink method. Few observatories can provide the long-term access to both spectroscopic and photometric facilities which is essential to projects of this kind. As a by-product of this work, Evans began to investigate the binary nature of some Cepheids, by searching for cycle-to-cycle variations in their mean radial velocities. Since binary Cepheids have orbital periods of one to ten years or more, regular access to a spectroscopic telescope is obviously necessary!

Variables in the Southern Skies

The 0.6 m telescope at Las Campanas offers many advantages for the study of stars in general. The clear, dark skies make it possible to reach magnitude 15 photoelectrically, faint enough to reach many of the variables in the Magellanic Clouds. The superb seeing results in very sharp star images, which is a great advantage in both imaging and spectroscopic studies. The use of image tubes further increases the power of the telescope, and this enables the telescope/spectrograph system to reach as faint as magnitude 14.

Most observing runs at Las Campanas are about three weeks in length. This restricts to some extent the kind of variables which can be studied, but southern Cepheids have been extensively observed, especially by Barry F. Madore (1975). Even Mira variables have been effectively studied from this site. Richard A. Crowe, while he was the resident observer at Las Campanas, accumulated a two-year supply of spectra of these variables, which later formed the basis of his doctoral thesis research. This study provided an effective complement to the study of northern Mira variables carried out by Robert F. Garrison and his colleagues at the Mt. Wilson Observatory many years ago.

Much of the observing time at Las Campanas is devoted to spectroscopy at "classification dispersion" (about 120 Å/mm), and the massive spectroscopic surveys by Robert F. Garrison are probably the most extensive surveys of their kind in the southern sky. They provide an invaluable body of basic data on stars, especially when combined with *UBV* data, and they have turned up dozens of variable or potentially variable stars which are worthy of further study. One notable product of these surveys was the discovery of CPD -48 1577, the brightest known cataclysmic variable. Variable star research at Las Campanas is described in more detail in the article by Garrison, elsewhere in this volume.

Epilogue

Fifty years ago, the astronomers at the DDO set out to undertake a few fundamental projects which were important and well-suited to the facilities available. Today, a larger and more diverse staff investigates a much wider set of problems with facilities spread around the Earth — and even in space. We still like to think of the DDO as "home base", not only in terms of the facilities located there, but also in terms of the origin of our scientific traditions — ones which unite staff and graduate students in the pursuit of solutions to fundamental problems in astronomy and astrophysics.

REFERENCES

Barnes, T.G. (1973). *Ap. J. Suppl.*, **25**, 369.
Bolton, C.T. (1972). *Nature Phys. Sci.*, **235**, 271.
Bolton, C.T. (1982). In *Be Stars*, IAU Symposium No. 98, ed. M. Jaschek & H.-G. Groth, p. 181. Dordrecht: D. Reidel.
Copeland, J.A. & Heard, J.F. (1963). *Pub. David Dunlap Observatory*, **2**, 315.
Demers, S. & Fernie, J.D. (1966). *Ap. J.*, **144**, 440.
DuPuy, D.L. (1973). *Ap. J.*, **185**, 597.
Evans, N.R. (1970). *A. J.*, **75**, 636.
Evans, N.R. (1976a). *Ap. J.*, **209**, 135.
Evans, N.R. (1976b). *Ap. J. Suppl.*, **32**, 399.
Fernie, J.D. (editor). (1973). *Variable Stars in Globular Clusters and in Related Systems*, Dordrecht: D. Reidel.
Fernie, J.D., Sherwood, V. & DuPuy, D.L. (1972). *Ap. J.*, **172**, 383.
Hall, D.S. (1976). In *Multiple Periodic Variable Stars*, ed. W.S. Fitch, p. 287. Dordrecht: D. Reidel.
Herbst, W. (1973). *Astr. Ap.*, **26**, 137.
Madore, B.F. (1975). *Ap. J. Suppl.*, **29**, 219.
Tanner, R.W. (1948). *J. R. A. S. Canada*, **42**, 177.

I. VISUAL OBSERVATIONS

Janet A. Mattei has been Director of the AAVSO for more than ten years, and is in a unique position to understand the nature of visual observations, and their role in astronomy. She is responsible for the coordination and archiving of millions of such observations; she is also aware of the enjoyment and satisfaction which variable star observing brings to those who do it. As a professional astronomer, she knows the scientific value of visual observations, both from her own research and from that of the hundreds of other professional astronomers who have requested AAVSO visual data in the past.

It is interesting to note that, as astronomy has expanded into the radio, infrared, ultraviolet and X-ray regions of the spectrum, the importance of visual observations has actually increased. A case in point is CH Cygni, a symbiotic binary star. Ernest R. Seaquist has discovered interesting and unpredicted behaviour of this star, using the VLA (Very Large Array) radio observatory in New Mexico. In his paper, he shows how long-term visual monitoring of this star has helped to illuminate its nature. There must be hundreds (or thousands) of other stars which are sufficiently interesting to deserve long-term monitoring, especially if they are also being studied at radio and other wavelengths.

Among the many types of variables described by Janet Mattei are the eclipsing variables. David B. Williams is an active AAVSO observer who has a special interest in these stars. In his paper, he illustrates the particular techniques and benefits of observing these. His light curve of V1010 Ophiuchi is a beautiful example of the precision which can be attained in visual observations if all factors are optimized.

VISUAL OBSERVING OF VARIABLE STARS

Janet Akyüz Mattei
American Association of Variable Star Observers (AAVSO)
25 Birch Street
Cambridge, MA 02138
U.S.A.

INTRODUCTION

There are over 28,000 stars known to be changing in brightness and another 14,000 stars suspected to be changing in brightness. These known and suspected variable stars need continuous, systematic observing over decades to determine their short-term and long-term behaviour, and to catch and record any unusual activity. During the last decade variable stars have been closely monitored using specialized instruments on ground-based large telescopes and x-ray, far-ultraviolet, ultraviolet, and infrared detectors aboard satellites. It is essential to have continuous optical data to correlate with multi-wavelength observations that have been obtained with these specialized instruments. The easiest and the most accessible method of monitoring variable stars is through visual observing. A serious observer can contribute valuable data which helps to increase our knowledge of these and other stars, our galaxy, and the universe at large.

The importance of the contribution of the serious amateur observer was first realized 141 years ago by Friedrich Wilhelm August Argeländer (1799 - 1875), a German astronomer famous for his *Bonner Durchmusterung* star atlas and catalogue, who is considered to be the father of variable star astronomy. In 1844 when only 30 variable stars were known, Argeländer wrote in an article: "... I lay these hitherto sorely neglected variables most pressingly on the heart of all lovers of the starry heavens. May you increase your enjoyment by combining the useful and the pleasant while you perform an important part towards the increase of human knowledge." Argeländer's plea is even more appropriate for today, with over 28,000 known variable stars in need of monitoring.

TYPES OF VARIABLE STARS

Variable stars are divided into four main classes: *pulsating* and *eruptive variables* in which variability is intrinsic — due to physical changes in the star or stellar system; *eclipsing binary* and *rotating stars* in which variability is extrinsic — due to an eclipse of one star by another or the effect of stellar rotation. A brief and general description of the major types in each class is given below, with accompanying visual light curves of some of them:

Pulsating Variables

Cepheids. Variables that pulsate with periods from 1 to 70 days, with an amplitude of light variation from 0.1 to 2 magnitudes. They have high luminosity and are of F spectral class at maximum and G to K at minimum. The later the spectral class of a Cepheid, the longer is its period. Cepheids also obey the well-known period-luminosity relation. Example: δ Cephei.

RR Lyrae stars. These are short-period pulsating variables with periods ranging from 0.05 to 1.2 days and light amplitudes between 0.3 and 2 magnitudes. They are white giant stars, usually of spectral class A. Example: RR Lyrae.

RV Tauri stars. These are yellow supergiants having a characteristic light variation with alternating deep and shallow minima. Their periods, defined as the interval between two deep minima, range from 30 to 150 days. The light amplitude may be as much as three magnitudes. Some show long-term cyclic variations from hundreds to thousands of days. Generally, the spectral classes range from G to K. Example: R Scuti.

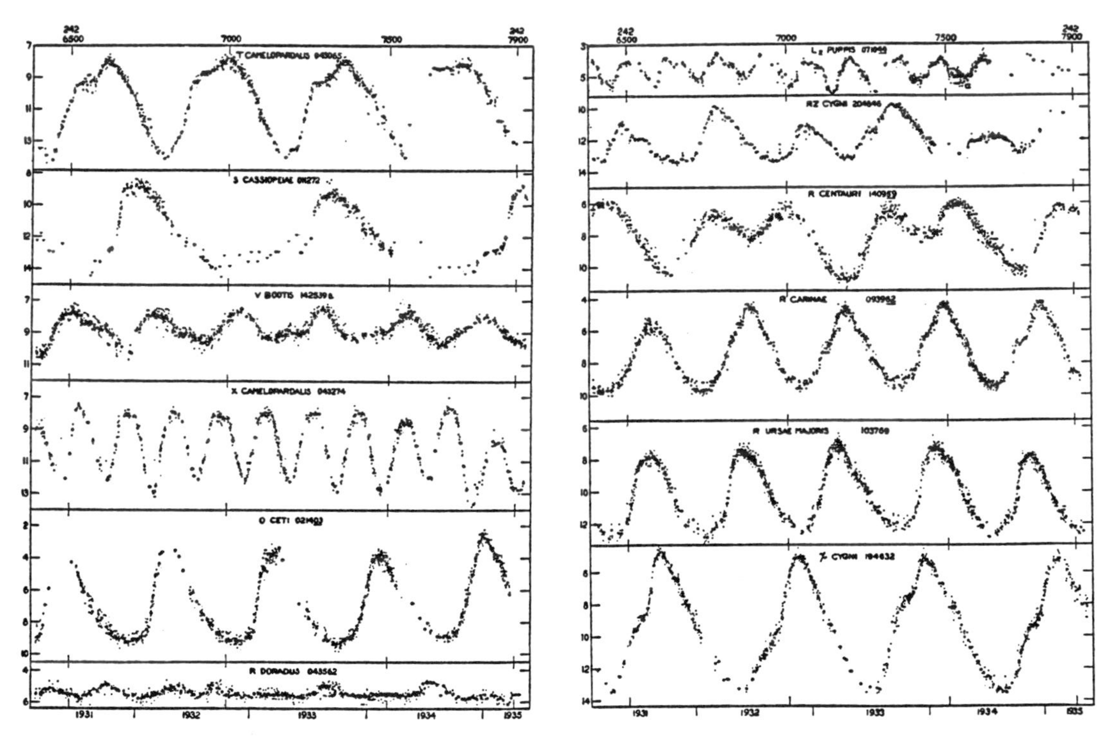

Figure 1. AAVSO light curves of 12 long-period (Mira) and semiregular variables, all plotted on the same scale.

Long-period or Mira variables. These are giant red variables that vary with periods ranging from 80 to 1000 days and visual light amplitudes from 2.5 to 5 magnitudes or more. They show characteristic emission, and spectral classes M, C, and S. Example: *o* Ceti (Mira).

Semiregular variables. These are giants and supergiants showing appreciable periodicity accompanied by intervals of irregular light variation. The periods range from 30 to 1000 days, with light amplitudes not more than one to two magnitudes. Example: Z Ursae Majoris.

Eruptive Variables

Supernovae. These show brightness increases of 20 or more magnitudes as a result of a catastrophic stellar explosion. Example: CM Tauri (Supernova of A.D. 1054 and the central star of the Crab Nebula).

Novae. These close binary systems, consisting of a normal star and a white dwarf, increase in brightness by 7 to 16 magnitudes in a matter of one to several hundred days. After the outburst, the star fades slowly to the initial brightness over several years or decades. Near maximum brightness, the spectra are generally similar to A or F giant stars. Example: CP Puppis (Nova 1942).

Recurrent novae. These objects are similar to novae, but they have undergone two or more outbursts during their recorded history. Example: RS Ophiuchi.

U Geminorum stars. These are close binary systems with periods on the order of a few hours. After intervals of quiescence at minimum light, they brighten suddenly. Depending upon the star, the eruptions occur at intervals of 10 to thousands of days. The light amplitude of the outbursts ranges from two to six magnitudes, the duration from 5 to 20 days. Example: U Geminorum.

Z Camelopardalis stars. These stars are physically and spectroscopically similar to U Geminorum stars. They show cyclic variations interrupted by intervals of constant brightness (stillstands). These stillstands last the equivalent of several cycles, with the star "stuck" at a brightness approximately a third of the way from maximum to minimum. Example: Z Camelopardalis.

SU Ursae Majoris stars. These stars are also physically and spectroscopically similar to U Geminorum stars. They have two distinct kinds of outbursts: one is short with a duration of one to two days, faint and more frequent; the other (supermaximum) is long with a duration of 10 to 20 days, bright and less frequent. During supermaxima, small-amplitude, periodic modulations (superhumps) appear

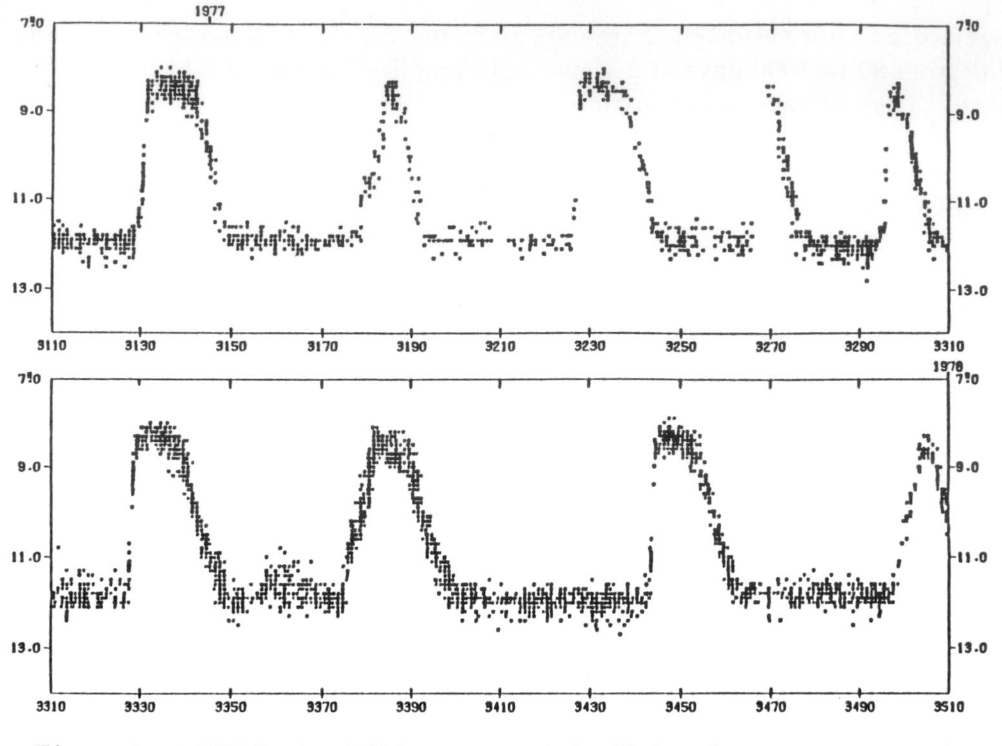

Figure 2. AAVSO visual light curves of the U Geminorum-type eruptive variable SS Cygni.

with periods two to three per cent longer than the orbital period of the system. Example: SU Ursae Majoris.

Symbiotic stars. These are close binary stars with one component a red giant, the other a hot blue star, and both embedded in nebulosity. They show semi-periodic, nova-like outbursts up to three magnitudes in amplitude. Example: Z Andromedae.

R Coronae Borealis stars. These are luminous, hydrogen-poor, carbon-rich, rare variables that spend most of their time at maximum light and at irregular intervals fade as much as nine magnitudes, then slowly recover to their normal brightness after a few months to a year. The drop in brightness is believed to be caused by the formation of "carbon soot" in the atmosphere of the star. Members of this group have F to K and R spectral types. Example: R Coronae Borealis.

Eclipsing Binaries

These are binary systems of stars with an orbital plane lying near the line of sight of the observer. The components periodically eclipse each other, causing

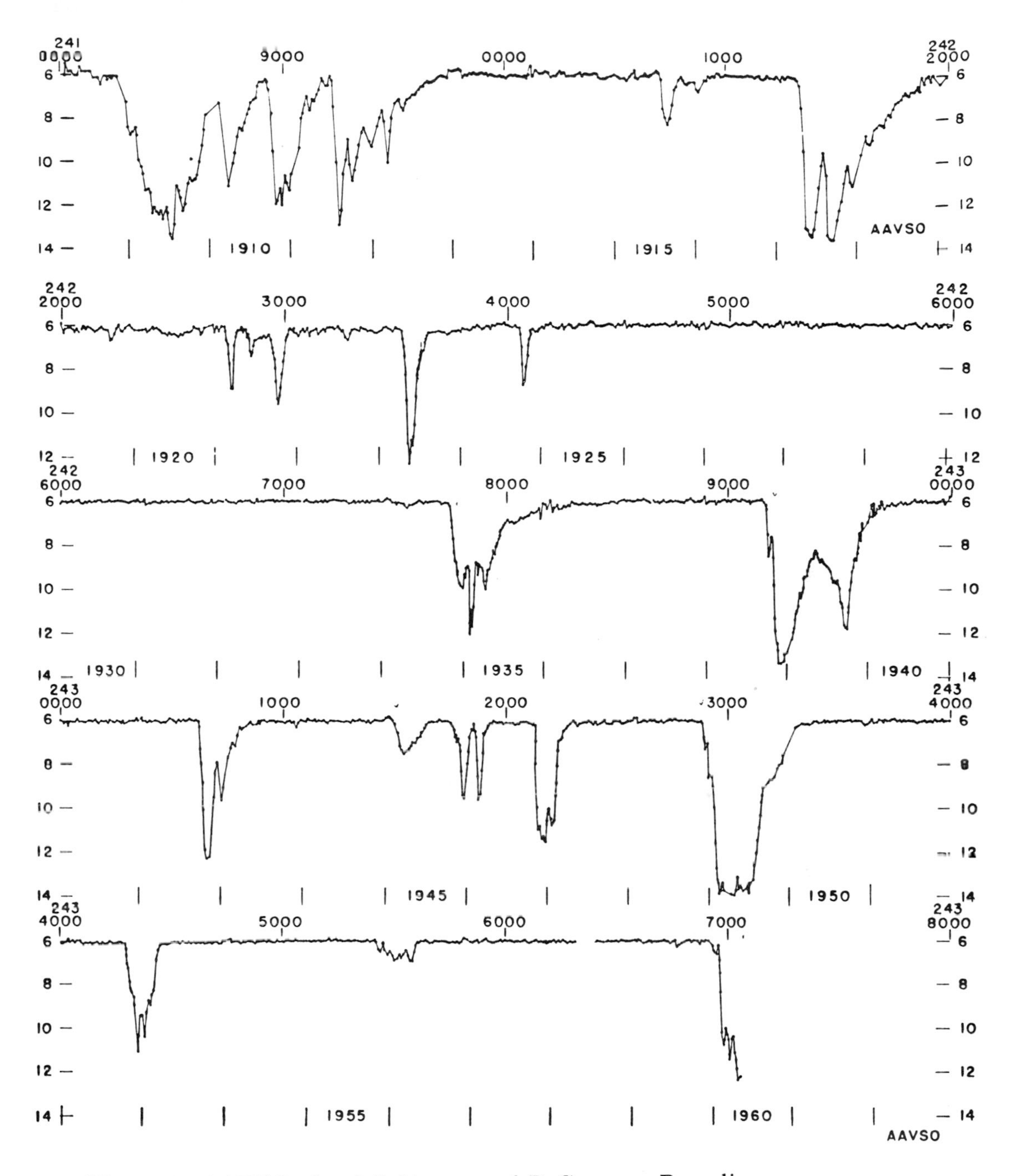

Figure 3. AAVSO visual light curve of R Coronae Borealis

a decrease in the apparent brightness of the system as seen by the observer. The period of the eclipse, which coincides with the orbital period of the system, can range from minutes to years. Example: β Persei (Algol).

Rotating Variables

These are rotating stars, often binary systems, which undergo small amplitude changes in light that may be due to dark or bright spots or patches on their stellar surface. Eclipses may also be present in such systems. Example: RS Canum Venaticorum.

From this summary of the major types of variable stars, those types that are best suited for visual observing have amplitudes of variation larger than one magnitude ($\Delta m \geq 1$). The best candidates are long-period and semiregular variables, R Coronae Borealis stars, RV Tauri stars, large- amplitude Cepheids and RR Lyrae stars, cataclysmic variables (novae, U Geminorum, Z Camelopardalis, SU Ursae Majoris, and symbiotic stars), and all the eclipsing binaries with amplitude of minima of one magnitude or more.

PREPARING A VISUAL OBSERVING PROGRAM

There are several points to consider in preparing a visual observing program. These are:

1. *Geographic Location.* The scale of the observing program will be influenced by the location and the terrain of the observer's site.

2. *Sky Condition.* The annual percentage of clear nights influences the selection of types of variables. The larger the percentage of clear nights the more advisable it is to go after stars that require nightly observations, such as the cataclysmic variables and R Coronae Borealis stars. If an observer has only 20% or less of clear nights per year, however, it is recommended that slowly varying long-period variables be selected for which even one observation a month is meaningful.

3. *Light Pollution.* Light pollution affects the limiting magnitude of observations. Thus, an observer living in a city is advised to concentrate on observing bright stars well above the limiting magnitude of the instrument, while observers with dark skies should be challenged to go after stars as faint as their instruments will allow.

4. *Optical Instrument.* Successful variable star observing requires interest, perseverance, and the proper optical instrument. A good pair of binoculars is sufficient for bright stars, while for faint stars one needs either a small or a large reflector or refractor telescope, either portable or permanently mounted. There is no "ideal" telescope for variable star observing; each has its own special advantage. More important is familiarity with the instrument at hand and a determination to accomplish as much as possible with it.

   The most popular type of telescope among variable star observers is a short-focus (f/5 - f/8) Newtonian reflector, with aperture six inches or more, equipped with a good finder. Such an instrument is easy to build or inexpensive to buy, and more important, it is convenient to use and easy to maneuver.

   A wide-field eyepiece is an important aid in locating variable stars, and it allows the observer to include as many of the comparison stars in the field as possible. High magnification is not necessary unless one is observing faint stars or congested fields. Recommended eyepieces are the 15 - 20 mm focal-length Kellner, 10 mm Kellner or orthoscopic, and 6 mm orthoscopic. A good quality, achromatic, two-power Barlow lens is also a valuable aid.

DEVELOPING AN OBSERVING PROGRAM

Developing a good, productive, and satisfying program should be an important goal for every observer. Generally, long-period and semiregular variables are recommended for beginners. These stars have a wide range of variation, are relatively easy to observe, and many of them are close to bright stars which are a help in locating them. Since these stars vary slowly, observing them three to four times a month is sufficient. After experience is gained observing these stars, the program can be expanded by adding other types of variables such as cataclysmic variables, symbiotic stars, and R Coronae Borealis stars, all of which exhibit unpredictable behaviour and require nightly observations.

The AAVSO helps set up numerous observing programs for new observers and individuals in universities, colleges, high schools, and astronomy clubs. AAVSO observing aids, such as an instruction manual, finding charts for 2000 variable stars, and the publications described in the accompanying list, are available to interested observers.

Also available is a customized AAVSO Observing Kit which includes a specially selected set of finding charts appropriate for the observer's interest, goals, experience, instrument, and location, together with detailed instructions for perparing an observing program.

SPECIAL OBSERVING PROGRAMS

Observing Eclipsing Binaries and RR Lyrae Stars

Once sufficient experience is attained, observers may wish to try their hand at using special observing techniques on such interesting objects as eclipsing binaries and RR Lyrae stars. For these stars the primary purpose of visual observing is to determine times of minima and maxima, respectively. Changes in the period of these stars can be detected by compiling an accurate record of times of minima and maxima over many years. But planning is essential for successful observations of short-period variables. The observer must select the observable stars, where the choice is governed by available instrument, location, and the time of the year. Little can be accomplished in observing eclipsing binary and RR Lyrae stars unless at least two hours can be devoted to each observing session.

To observe minima of eclipsing binaries it is advisable to locate the star field a few nights in advance, and to examine all comparison stars in the field. Then, on eclipse night, plan on observing early enough to obtain the descending portion of the light curve. Continue the observations after mid-eclipse to obtain the ascending portion as well. A magnitude estimate every ten minutes is recommended, although the frequency of observations will depend upon the rapidity of the light variation. All observations should be timed to the nearest minute, with an accurate clock. A most common pitfall for short-period variable star observers is that of bias introduced by anticipation. Never let the results of a previous observation influence the present one. Record what you see and not what you think you should. In timing maxima of RR Lyrae stars a procedure similar to that given for eclipsing binaries is recommended.

Searching for Novae and Supernovae

A quite different type of variable star observing involves searching for exploding stars, or novae. In the AAVSO Nova Search program, the regions of the Milky Way where novae are most likely to occur have been divided into areas 10 degrees in declination, by about one hour in right ascension. Observers are assigned specific areas and are advised to memorize star patterns and check these areas nightly for novae. The standard equipment for a nova search is a good star atlas and a pair of 7×50 binoculars.

Some observers may be interested in searching galaxies for supernovae. A telescope with a limiting magnitude of at least 13, and a set of good finding charts of galaxies are essential. Discovering a supernova requires extreme familiarity with the galaxy and the surrounding field stars, and care in observing should be exercised. In this century, until 1980, there had been only two visual discoveries of supernovae. This record has been spectacularly broken by Reverend Robert Evans of Australia who with his perseverance, meticulous observing, good finding charts, and photographs of galaxies, has made 14 visual discoveries of supernovae. Reverend Evans' success is remarkable, indeed!

MAKING OBSERVATIONS

The AAVSO observing program is based on sets of standardized finding charts for all the variables that are in the program. We encourage our observers to use these charts to avoid as much as possible the conflict that can arise when magnitudes for the same comparison star are derived from different sets of charts, resulting in two different degrees of variation being recorded for the same star.

A detailed observing log is the best way of ensuring that the quality of the observation is as much a part of the record as the magnitude estimate itself. Case in point: some years ago, we had two sets of observations made by two excellent observers, one reporting a cataclysmic variable at minimum with the other observer reporting the star at maximum, at the *same* time. When asked to confirm the quality of their observations, one of the observers responded: "I am sure that the star was at minimum on that date. My observing log does not indicate any uncertainty with that observation. I used the following comparison stars ... the sky was clear, there was no moonlight, and so I can't fault my data." The other observer responded: "I don't know what I was doing that night; that observation is not good." So it is important that *everything* be written down, with the observations properly logged, since one can never tell when an observations may be questioned.

At the start of an observing session, record the date (preferably unambiguously, *e.g.*, April 8-9/86), and the sky conditions — particularly the transparency, cloud cover, presence of moonlight, and any change in observing conditions. Make a magnitude estimate of the variable star by comparing its brightness with a brighter and fainter comparison star in the field. Then record your estimate, the comparison stars used, and the exact time of your observation.

We ask our observers to report at least the following: the star's designation (given on the finding chart) which is the approximate position of the variable (the first four figures are the right ascension, and the last two figures are the declination); the name of the variable; the date (in Julian Day), the time of observation to the minute, and the kind of time that is used (*i.e.*, Eastern Standard Time, Greenwich Mean Astronomical Time, Universal Time); the magnitude estimate of the variable and the magnitudes of the comparison stars used. The AAVSO provides tables for calculation of Greenwich Mean Astronomical Time, as well as tables to determine the Julian Day.

No matter what you observe, or when, or for whom, always submit or publish your observations as a matter of record. Observations that remain forgotten in a desk drawer are of no value to science.

SCIENCE WITH VISUAL DATA

The accuracy of individual visual observations is generally ±0.4 and at best ±0.2 magnitude. Therefore visual variable star data containing compiled observations of many observers are best suited for studies of long-term, large-amplitude ($\Delta m \geq 1$) behaviour of variable stars. Studies using compiled visual data of stars with less than one magnitude of variation range are stretching these observations beyond their limit, and thus the results may be questionable.

AAVSO receives an increasing number of requests for its visual data from astronomers each year. Figure 4 is a histogram of the number of requests for data fulfilled each year since 1974. The requests are for data on all types of variable stars, but due to very active satellite research programs in recent years, the largest number of requests have been for data on cataclysmic variables. Figure 5 is a piegram showing the different types of variable stars for which AAVSO data have been requested in the last year (Mattei 1984).

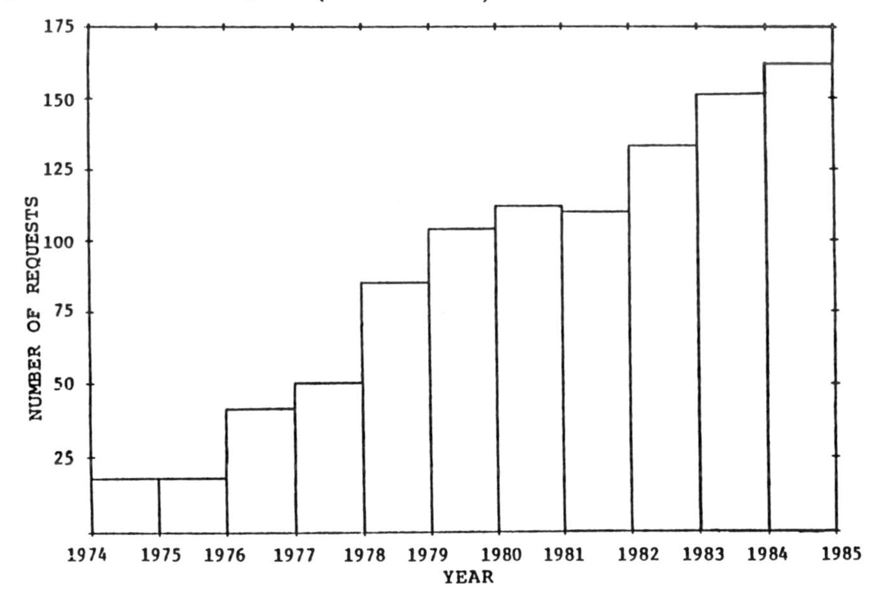

Figure 4. Number of special requests for AAVSO visual data received and filled each year since 1974.

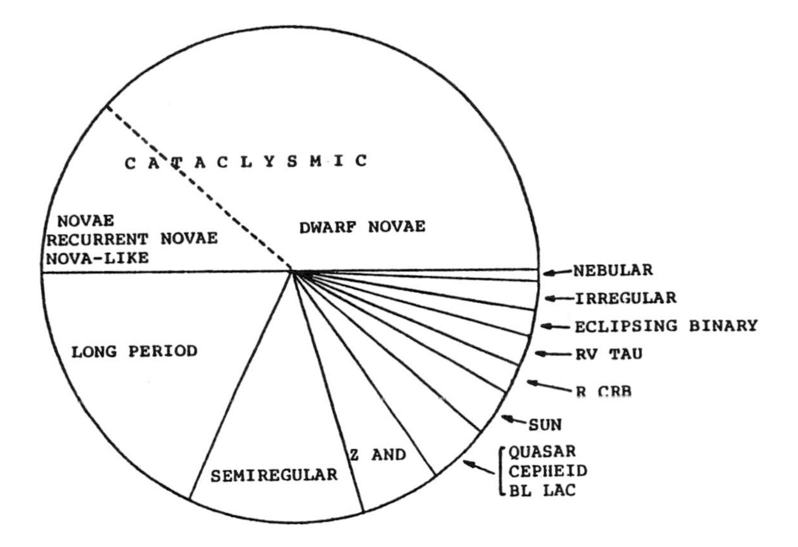

Figure 5. Types of variable stars for which data were requested in 1984.

The AAVSO visual data have been used in the following areas:

1. *Data Analysis.* AAVSO visual data have been used extensively to analyze the long-term behaviour of long-period, semiregular, R Coronae Borealis, and cataclysmic variables; to determine maxima and minima and to revise the periods of long-period and semiregular variables, Cepheids, RR Lyrae stars, and eclipsing binaries; and to correlate different parameters of the light curves.

2. *Data Correlation.* Visual data have been used to correlate photometric behaviour with spectroscopic data made in the radio, infrared, ultraviolet, far ultraviolet, extreme ultraviolet, and x-ray wavelengths using large ground-based telescopes at observatories such as Kitt Peak, Mount Palomar, Cerro Tololo, Lick, McDonald, Mount Lemmon, Whipple, Dominion Astrophysical, Very Large Array Radio Telescope, and instruments aboard satellites such as the International Ultraviolet Explorer, High Energy Astronomical Observatory 1 and 2, Apollo-Soyuz, Voyager, Infrared Astronomical Satellite, and the European X-ray Observatory Satellite.

3. *Scheduling Observing Runs, and Real-Time Notification of the Behaviour of Observing Targets.* Continuous visual observations are of great help to astronomers who wish to schedule observing runs during a specific phase of the light curve, with large ground-based telescopes or orbiting satellites equipped with specialized instruments. Using AAVSO's long-term and up-to-date data files, we are able to predict the date and the brightness of the

variable at the requested phase. Then during the observing run, through real-time participation of visual observers, the interested astronomers are kept continuously informed on the optical behaviour of these stars. This type of information has been crucial for the most effective use of telescope and instrument time, and for the successful execution of these observing runs. Just in the last year alone AAVSO has helped to schedule 36 observing runs, and has provided real-time information for 20 observing runs using satellites.

4. *Theoretical Research.* Theoretical models of variable stars are valid only if they can be fitted to observations. Long-term data of the AAVSO has played a crucial role in checking and confirming theories of these stars.

STUDIES WITH VARIABLE STARS

The period of variation — the time measured between two consecutive maxima or minima of a variable star — is a very important parameter. Depending upon the type, the period can be an indicator of the size, luminosity, age, and mass of a star. Determining an accurate period of a star requires decades of monitoring many cycles. Quite often the period of a variable star will not remain constant. Changes in the period of a system may be an indication of changes in the physical structure and evolution of the system. The simplest and best means of examining the constancy of the period of a star is by obtaining its times of minima (for eclipsing binaries) or maxima (for other types of stars) and constructing an $O - C$ diagram (see article by Willson elsewhere in this volume).

In the following section, I will give some specific examples of how the visual observations have been used in the study of eclipsing binaries, and pulsating and eruptive variables.

Studies with Eclipsing Binary Stars

Visual observations of minima of eclipsing binaries have provided a wealth of information on the times of minima of these stars. This information is mainly used to search for period changes, through the construction of an $O - C$ diagram, which may then indicate physical and evolutionary changes in the system. The new revised elements obtained from such studies can then help to predict future times of minima. The result of one such study using visual observations of the eclipsing binary X Trianguli is shown in Figure 6. The eclipse period of this star is given as 0.9715382 days in the third edition of the *General Catalog of Variable Stars.* However, $O - C$ residuals plotted against cycle from 1965 to 1974 indicate that the period has decreased twice, with a decrease to 0.9715270 days in recent years as seen from the steeper slope (Baldwin 1974).

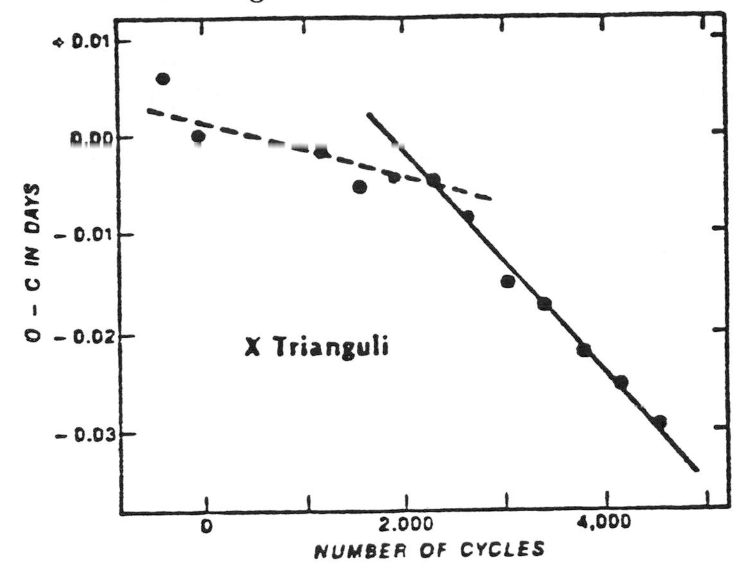

Figure 6. Graph of the $O - C$ residuals vs. cycle from 1965 to 1974 for the eclipsing binary X Trianguli. Note the abrupt change in period halfway through the time interval.

Studies with Pulsating Variable Stars

The largest portion of the AAVSO observing program is devoted to red giant pulsating variables — the Mira stars — and to semiregular variables, for which homogeneous and continuous data have been compiled for decades. Light curves of 12 long-period variables (Figure 1) indicate that, within these periodic pulsating variables, there exists significant light variation in each star, and, within each star, variations exist from cycle to cycle. Therefore, continuous monitoring is essential in understanding these systems.

Valuable information may be obtained by studying the possible correlations between parameters of the light curves (period, amplitude, spectra and rates of rise to maximum and decline to minimum). Currently a joint research project of the University of Toronto, AAVSO, and the Iowa State University has been to obtain $O - C$ curves for about 350 long-period variables, using 75 years of maxima and minima dates from AAVSO data files. The aim of the project is to search for period changes and to find relationships between various parameters of the light curves with the goal of trying to understand these stars better.

Recent AAVSO visual data have been extensively used to determine the phase and the brightness of pulsating variables during multicolour photometric or spectroscopic observations, and to correlate these results with the light variation. One such interesting study involved a Mira variable, R Cygni. This variable has a mean period of 429 days, and a mean visual brightness at maximum of 7.5 magnitudes. Interestingly, R Cygni has significant brightness variations at maximum as shown in the graph in Figure 7. A study of the long-term light curve reveals a correlation between brightness at maximum and the time interval since the previous maximum, in that fainter maxima occur later than normal and are followed by maxima that occur earlier than normal. Spectroscopically, the emission and absorption lines in the optical and near infrared reveal significant correlations as well. The velocities of the emission lines correlate with the magnitude at maximum, in that during bright maxima they are negatively displaced with respect to the red component of absorption lines, while during the faintest maximum there is no displacement. This has been interpreted as possibly due to a reduced velocity of expansion during faint maximum, or an enhanced degree of limb-brightening during low maxima (Wallerstein *et al.* 1985).

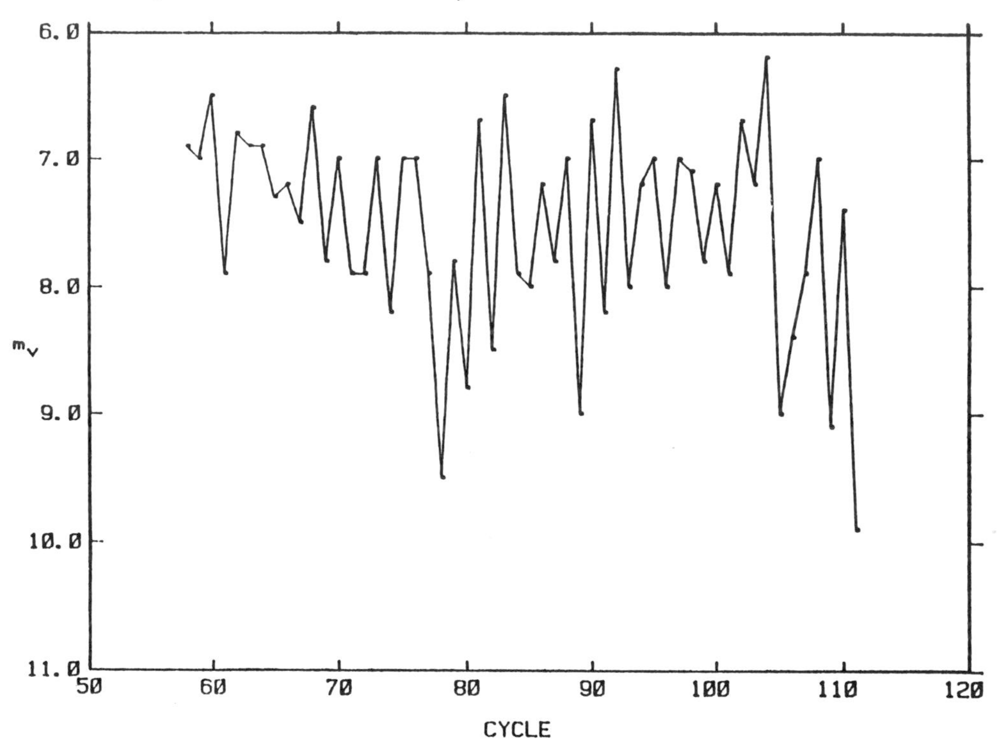

Figure 7. Graph of the visual magnitudes at maximum vs. cycle count from 1921 to 1983 of the long-period variable R Cygni.

In the area of semiregular variables, visual data have been used extensively to study their regularity and to search for multi-periodicity in these stars. In an interesting project in progress involving the AAVSO and Grinnell College, we have studied the long-term light curves of some well-observed semiregular variables and have determined which ones have extended intervals of inactivity. During their inactive times these stars are monitored photometrically at Grinnell College to determine if they show small amplitude periodic variations that are otherwise not detectable through visual observing. The findings of this project will have a bearing on our understanding of the pulsation mechanism of these stars and how they are related to the periodic and more regular Mira variables.

Studies with Eruptive Variables

A study of long-term data can yield much valuable information about the close binary eruptive stars, often referred to as cataclysmic variables. Light curves of these stars contain invaluable information on their eruptions (outbursts), the nature and origin of which is still not well understood. The data on these stars are doubly important since an understanding of these systems can contribute to a better understanding of other wide-ranging astrophysical issues. These issues include time-varying processes in eruptive variables (recurrence times, duration, differing rates of rise and decline from outbursts, unusual changes in the pattern of events), the structure of the complex binary system (the cool main sequence star and its accreting blue companion, the white dwarf), the dynamics of accretion, and the physics of accretion disks, the evolution of these systems, and their relation to other non-eruptive close binary systems.

The analysis of long-term visual data on cataclysmic variables has thus far revealed interesting correlations. There is, for example, a strong correlation among the duration of the outbust, rate of rise and decline from outburst and the orbital period and the mass of the secondary component of the system (Szkody & Mattei 1984). There is a relation between the orbital period and the rate of decline from outburst, in the sense that the smaller the orbital period the faster is the decline from outburst (Bailey 1975; Mattei & Klavetter 1981). The orbital period of an eruptive close binary system is an important physical parameter that has a direct bearing on the mass of the secondary, the rate of mass transfer, the size of the system, and possibly the evolution of the system. An indirect way to obtain the orbital period of a special type of cataclysmic variable — SU Ursae Majoris stars — is to observe the small-amplitude, periodic modulations (superhumps) during long outbursts (supermaxima) and to obtain their period — which is two to three per cent longer than the orbital period of the system itself. Thus, if supermaxima of a system can be predicted, then through the observations of superhumps the orbital period of an SU UMa system may be determined. Long-term analysis of the light curves of these stars has shown that one can predict the occurrence of supermaxima (Mattei 1983) and several orbital periods have been obtained in this way (Bond *et al.* 1982).

Cataclysmic variables have been of great interest in high energy astrophysics studies because of their emission at x-ray, extreme ultraviolet and ultraviolet wavelengths. Since the early 1970's, these stars have been the observing targets for instruments aboard x-ray and ultraviolet satellites. Visual observers have participated in these observations through their close monitoring and notification of activity in eruptive variables. These dedicated observers have played an important role in the first detection of x-ray pulsations of these stars, as well as the discovery that 70 per cent of these stars show non-phase-dependent x-ray emission (Cordova *et al.* 1981; Cordova & Mason 1984). This information has been vital in testing theories of high energy emission in these compact systems.

CONCLUSION

The AAVSO has compiled over 5,000,000 observations of variable stars since its founding in 1911. On over 500 stars, AAVSO data files go back 100 years or more, and include the observations of Friedrich Argeländer and other well-known 19th century variable star astronomers. Through these long-term data files, more recent observations, and real-time notification to the astronomical community of unusual behaviour of stars, visual observers are playing a vital and unique role in variable star astronomy. I would like to conclude my article with a collage of statements from letters of astronomers who have used the visual data of the AAVSO:

"... For my observations of red giant stars, the importance of the work of the AAVSO cannot be overstated. The difficulty in securing photographic spectra and photometric magnitudes simultaneously, and the irregularities of the periodic variables, make continuous monitoring of variable stars by AAVSO observers all the more necessary."

T. Ake
Space Telescope Science Institute

"... on our research flights on the Kuiper Airborne Observatory ... we have relied on the AAVSO data to both predict the apparent magnitudes at the time of our observations, and to determine the phase of our observations. We have always found the AAVSO data to be excellent and reliable."

B. T. Soifer
California Institute of Technology

"I have found AAVSO data to be of greatest use in my work ... predicted times of maxima and minima for long-period variables are indispensable for preparing observing programs."

G. Wallerstein
University of Washington

"... the AAVSO has several times recently provided me with data on the visual behaviour of variable stars being observed at other wavelengths by satellite or ground-based facilities .... In one classic case recently we obtained precise simultaneous observations of the dwarf nova SU UMa with EXOSAT, IUE, and IRAS (satellites). Although several large ground-based observatories were prepared to make optical observations they were all 'clouded out'. The only point in this part of the spectrum was obtained by an AAVSO member .... This is only one example of the many occasions that the AAVSO has provided invaluable assistance."

M. Bode
University of Manchester, England

"... I have found that the AAVSO is extremely important in the planning and execution of observations at large telescopes and with satellites at x-ray and UV wavelengths. The IUE satellite typically takes one hour to set up on a new object .... I cannot afford satellite time to move to an object in the wrong outburst state .... This is prevented by notification from the AAVSO network."

P. Szkody
University of Washington

"... the AAVSO data base is one of the most important being kept in astronomy today."

R. Gehrz
University of Minnesota

LIST OF AAVSO PUBLICATIONS FOR THE VISUAL OBSERVER

1. *The Journal of the AAVSO*, contains papers relevant to variable star research; Reports of the AAVSO Committees, Director, and Treasurer; Table of AAVSO observers' totals; Minutes of AAVSO Meetings by the Secretary; book reviews; letters to the Editor. Sent to all members. 2 issues per year. $20.00 U.S.A., Canada, Mexico; $24.00 elsewhere.

2. *AAVSO Bulletin*, contains the annual predictions of maxima and minima dates of long period variable stars, to assist in planning an observing program. The Bulletin is sent to all active AAVSO observers. Issued annually. $20.00 U.S.A., Canada, Mexico; $24.00 elsewhere.

3. *The AAVSO RR Lyrae Variable Star Ephemeris*, contains annual predictions of times of maxima for large amplitude RR Lyrae stars. Sent to active AAVSO observers. Issued annually. $5.00 U.S.A., Canada, Mexico; $6.00 elsewhere.

4. *The AAVSO Eclipsing Binary Ephemeris*, contains annual predictions of times of minima for large amplitude eclipsing binary stars. Sent to active AAVSO observers. Issued annually. $5.00 U.S.A., Canada, Mexico; $6.00 elsewhere.

5. *The RR Lyrae Bulletin* and *The Eclipsing Binary Bulletin*, inform observers of the unusual activity of these stars in the observing program and list the stars that need special observing. Sent to members upon request. Issued irregularly.

6. *The AAVSO Circular*, contains monthly preliminary observations of some eruptive and other interesting variables submitted to this publication and includes a list of stars in the observing program needing more observation. Issued monthly. $10.00 U.S.A., Canada, Mexico; $14.00 elsewhere.

7. *The AAVSO Alert Notice*, provides immediate information on the discovery of novae and extra-galactic supernovae, the unusual activity of variable stars in the AAVSO observing program, and requests from astronomers for AAVSO observers to participate in monitoring specific variables. Issued as needed. $5.00 per year for postage and handling.

8. *Newstar*, AAVSO Nova/Supernova Search Newsletter, contains news on the discovery of novae and supernovae and other helpful information on nova/supernova search. Sent to active nova/supernova searchers and members upon request.

9. *The AAVSO Reports*, contain computerized light curves of variable stars in the AAVSO observing program, usually covering an interval of 1000 days. Issued as completed. $30.00 U.S.A., Canada, Mexico; $34.00 elsewhere.

10. *The AAVSO Monograph Series*, contains long-term (20 years or more) computer-generated light curves of AAVSO observations contributed by members/observers, one star per Monograph. Issued as completed. Available for postage and handling fee.

REFERENCES

Bailey, J. (1975). *J. British Astron. Assoc.*, **86**, 30.

Baldwin, M.E. (1974). *J. Amer. Assoc. Var. Star Obs.*, **3**, 28.

Bond, H., Kemper, E., & Mattei, J.A. (1982). *A. J.*, **260**, 179.

Cordova, F., Jensen, K., & Nugent, J. (1981). The HEAO-1 Soft X-Ray Survey of Cataclysmic Variable Stars. *M. N. R. A. S.*, **196**, 1.

Cordova, F., & Mason, K. (1984). X-Ray Observations of a Large Sample of Cataclysmic Variable Stars using the Einstein Observatory. *M. N. R. A. S.*, **206**, 879.

Kukarkin, B.V. *et al.* (1969). *General Catalog of Variable Stars.* Moscow.

Mattei, J.A. (1982). *Bull. AAS*, **14**, 879.

Mattei, J.A. (1984). *Observers' Handbook of the Royal Astron. Soc. Canada.* Toronto.

Mattei, J.A. (1984). *J. Amer. Assoc. Var. Star Obs.*, **13**, 86.

Mattei, J.A. & Klavetter, J. (1981). Paper given at the UC Santa Cruz Summer Workshop in Astronomy and Astrophysics.

Mattei, J.A., Mayer, E.H., & Baldwin, M.E. (1980). *Sky and Telescope*, **80**, 285.

Szkody, P. & Mattei, J.A. (1984). *Pub. A. S. P.*, **96**, 988.

Wallerstein, G., Hinkle, K.H., Dominy, J.F., Mattei, J.A., Smith, V.V., & Oke, J.B. (1985). *M. N. R. A. S.*, **215**, 67.

THE EXPANDING RADIO JET IN CH CYGNI: A CASE OF COOPERATION BETWEEN RADIO ASTRONOMERS AND OBSERVERS WITH SMALL OPTICAL TELESCOPES

E. R. Seaquist
David Dunlap Observatory
Department of Astronomy
University of Toronto
Toronto, Ontario
Canada M5S 1A1

I describe briefly the discovery of a radio emitting jet associated with the 8th magnitude symbiotic-like star CH Cyg, and its relationship to the optical light curve prepared by the American Association of Variable Star Observers (AAVSO). Symbiotic stars are stars which show evidence of a hot component and a cool component; most are thought to be binary stars.

As part of a program to monitor radio emission from selected symbiotic stars, Dr. A. R. Taylor (University of Groningen) and I observed CH Cyg on several occasions since April 1984 using the Very Large Array (VLA) of the National Radio Astronomy Observatory*. Between April and November 1984, a remarkable increase in the flux density occurred, reaching 40 mJy by January 1985. Figure 1 shows VLA maps at 2 cm wavelength made on 1984 November 8 and 1985 January 22, showing that a jet was produced, expanding in both directions by a factor of two in 75 days. The expansion rate is 1.1 arcsec year^{-1} corresponding to a transverse velocity of 1100 km s^{-1} in each direction. The epoch of emergence of the jet is estimated to be between July 22 and Sept. 7, 1984. A detailed analysis of this jet by Taylor, Seaquist, & Mattei (1985) indicates that the emission mechanism is thermal bremsstrahlung from ionized gas whose mass exceeds 2.4×10^{-6} solar mass.

The AAVSO visual light curve for CH Cyg, kindly provided to us by Dr. Janet Mattei, is reproduced in Figure 2. It is clear that the epoch of emergence of the jet (July – September, 1984) coincided with a decline by $1.^m5$ in the visual magnitude. This decline ended a seven-year-long maximum in the light curve of CH Cyg. The coincidence in time of these two events will be an important consideration for the production of a physical model for the jet. Our current model for the phenomenon involves supercritical accretion from a disk surrounding a white dwarf companion to the red giant in the CH Cyg system. The change in the visual luminosity probably reflects a change of state in the luminous accretion disk.

---

* The National Radio Astronomy Observatory is operated by the Associated Universities Inc. under contract to the National Science Foundation.

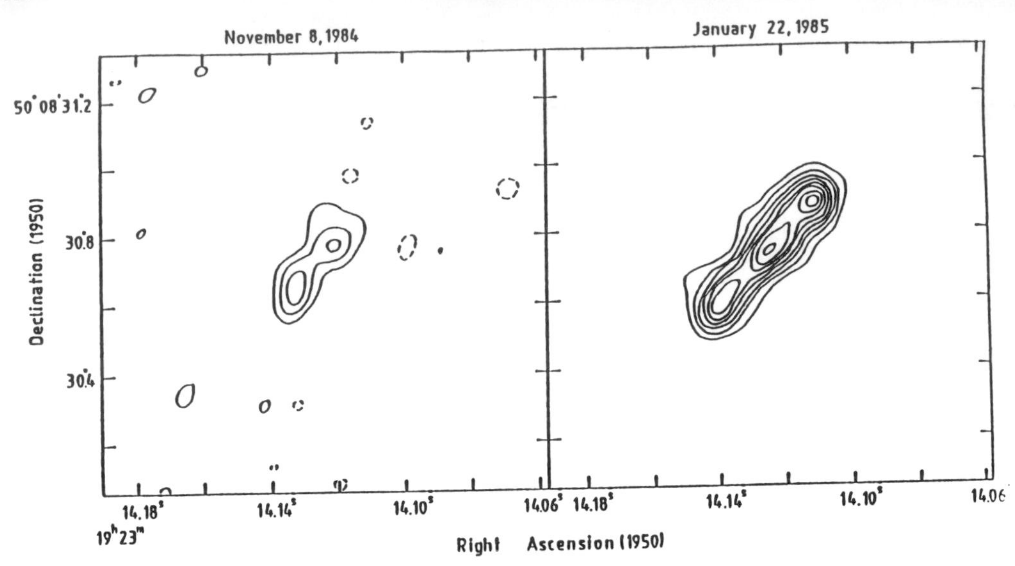

Figure 1. VLA radio maps at a wavelength of 2 cm showing the CH Cyg jet at two epochs.

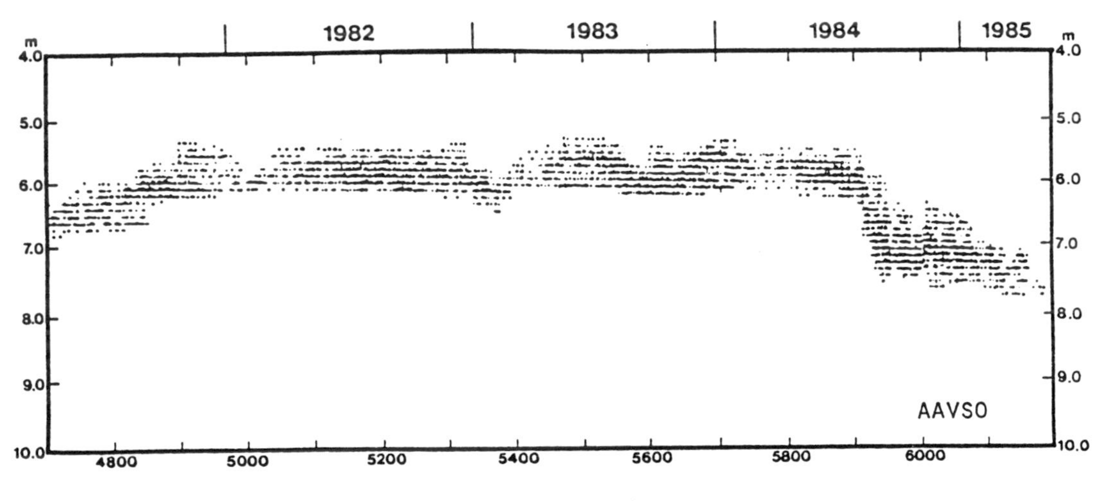

Figure 2. The AAVSO light curve for CH Cyg from April 1981 to May 1985. Each dot represents one observation. Reprinted by permission from Nature, Vol. **319***, No. 6048, pp. 38-41. Copyright ©1986 Macmillan Journals Limited.*

The data presented here provides an example of the importance of long term observing programs of variable stars by professional and amateur astronomers using smaller telescopes. A full understanding of the nature of many variable stars is possible only by observing these objects simultaneously in several regions of the electromagnetic spectrum. Campaigns to achieve this cooperation among large telescopes in different wavelength bands are frequently difficult or impossible to arrange. In many cases smaller telescopes can provide the complementary observations to enhance the value of observations in the radio, infrared, ultraviolet, or X-ray regions. Continuous optical coverage of the kind available for CH Cyg can generally be provided only by smaller telescopes. In the case of a serendipitous discovery such as the CH Cyg jet, it is the continuity of the optical coverage that is essential to maximize the benefits of this good fortune.

The description of this discovery is merely one example among many of the value of the long record of observations compiled by the AAVSO. As the period of overlap with VLA observations increases, the type of data provided by the AAVSO will play an increasingly important role in understanding radio-emitting stars.

REFERENCE

Taylor, A.R., Seaquist, E.R., & Mattei, J. (1986). *Nature*, **319**, 38.

VISUAL OBSERVATION OF ECLIPSING BINARY STARS: V1010 OPHIUCHI

David B. Williams
The American Association of Variable Star Observers

Traditionally, visual observation of variable stars has been concentrated on the intrinsic variables of the Mira, semiregular, and cataclysmic types. In the past twenty years, however, increasing attention has been given to the eclipsing binaries.

Many eclipsing binaries undergo slight and unpredictable changes in their orbital periods, usually as the result of mass exchange between the component stars or loss of mass from the system as a whole. More rarely, the apparent period will vary in a cyclical pattern that reveals orbital motion of the eclipsing pair around a third star. In other cases, the times of secondary minima may shift relative to primary minima, indicating rotation of the line of apsides in a binary with elliptical orbits.

Period changes in eclipsing binaries normally amount to no more than a few seconds in orbital periods which range from a third of a day to several days or more. A series of visual brightness estimates made while the variable is fading to minimum and then brightening can yield a time of mid-eclipse accurate to about ten minutes. This hardly seems a promising method for detecting period changes of a second or two. But any systematic change in the interval between eclipses is *cumulative* with each orbital revolution, and the observed times of mid-eclipse may eventually occur many minutes, or even hours, earlier or later than predicted by the original period. Visual observations over a number of years (not necessarily by the same observer) are therefore adequate to determine the periods in days of many eclipsing binaries to six or seven decimal places — a fraction of a second.

Monitoring the periods of eclipsing binaries, and revising their elements when necessary, is a very useful service to astronomers who investigate these stars with large and sophisticated instruments and must usually schedule their telescope time on the basis of the published elements.

Recently, for example, the International Ultraviolet Explorer satellite was used to obtain spectroscopic observations of V1010 Ophiuchi, a puzzling interacting eclipsing binary. If the observers had relied on the elements in the 1969 *General Catalog of Variable Stars*, they would have missed the critical phases of eclipse by $2\frac{1}{2}$ hours, and their allocation of valuable satellite time might have been wasted.

More than a thousand eclipsing binaries are observable visually with telescopes of 15 to 40 cm aperture. A few have no published period, and many have incorrect periods. Many exhibit eclipses that deviate by an hour or more from the published elements.

While greater precision can be achieved photoelectrically, one compensating advantage of visual work is the fact that the eclipses of several variables can be followed simultaneously. I have managed to observe the eclipses of up to eight stars at the same time, using binoculars and 10 cm and 20 cm reflectors. My current observing project on V1010 Ophiuchi illustrates the kind of results that are possible with visual techniques.

V1010 Ophiuchi (HR 6240, HD 151676, Sp.T. A5) is a 6th-magnitude β Lyrae-type eclipsing binary with a total range of $6.^m2 - 6.^m9$ and a brief interval of constant light at mid-eclipse. The variability was discovered as recently as 1964 at Bamberg Observatory. The next year, E. Schöffel and U. Köhler determined light elements that were adopted in the 1969 *General Catalog of Variable Stars*:

$$JD_{min} = 2425827.455 + 0.^d661436 \times E \qquad (1)$$

In 1965, I observed V1010 with 7×50 binoculars to determine whether it was suitable for visual timings of minima and to check the published period. I found that the primary minimum can be adequately defined by careful estimates at 15-minute intervals during an observing session of at least 3 hours centered on the time of mid-eclipse. Two times of primary minima agreed well with equation (1). During the same observing season, another AAVSO observer, Leonard Kalish, made photoelectric measures of nearby comparison stars in yellow light.

In 1982, I undertook new visual observations of V1010 to test the continued accuracy of equation (1) and to form a complete light curve. These observations revealed that minima were occurring almost three hours earlier than predicted by equation (1). I continued observations in 1983 and 1984, obtaining a total of 312 estimates using Kalish's photoelectric comparison star values.

These estimates were reduced to phase according to equation (1) and a composite light curve was plotted for each season. The phases of minima derived from these plots were then converted to $O - C$ values (see article by Willson in this volume) in thousandths of a day. In addition, eight individual times of minima were derived from single-night series of estimates.

A plot of these 11 $O - C$ residuals, and two published by a Swiss observer (Figure 1) indicates that the period of V1010 was $0.^d6614205$ during the interval of observation.

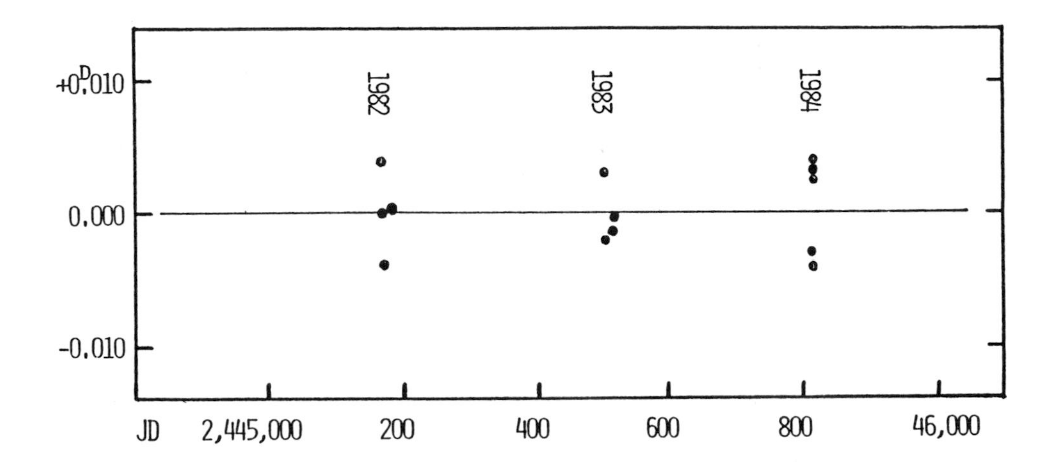

Figure 1. Observed-minus-Calculated plot for 11 times of primary minima observed by Williams and two by Maraziti. O − C residuals were calculated from the light elements

$$JD_{min} = 2445166.417 + 0.^{d}6614205 \times E$$

This revised period was used to form a single composite light curve from all 312 estimates (Figure 2). The normal points on this plot all fall within $0.^{m}05$ of a smooth curve. The good definition of the shallow secondary minimum demonstrates that visual estimates can produce useful results for stars varying by less than $0.^{m}5$ when the following conditions are met.

(1) The variable must be strictly periodic, permitting the averaging of observations over many cycles (this is the case in most eclipsing variables and Cepheids, and some RR Lyrae stars).

(2) Comparison stars must be conveniently placed, and differ by small but accurately known magnitude intervals. The total range of V1010 is well-matched by nearby comparison stars of $6.^{m}12$, $6.^{m}49$, and $7.^{m}16$. The intermediate star closely approximates V1010's brightness at secondary minimum, which permits the careful observer to define this portion of the light curve particularly well.

(3) All observations must be made by the same observer with the same instrument in a consistent manner. Even so, the observer's "personal equation"

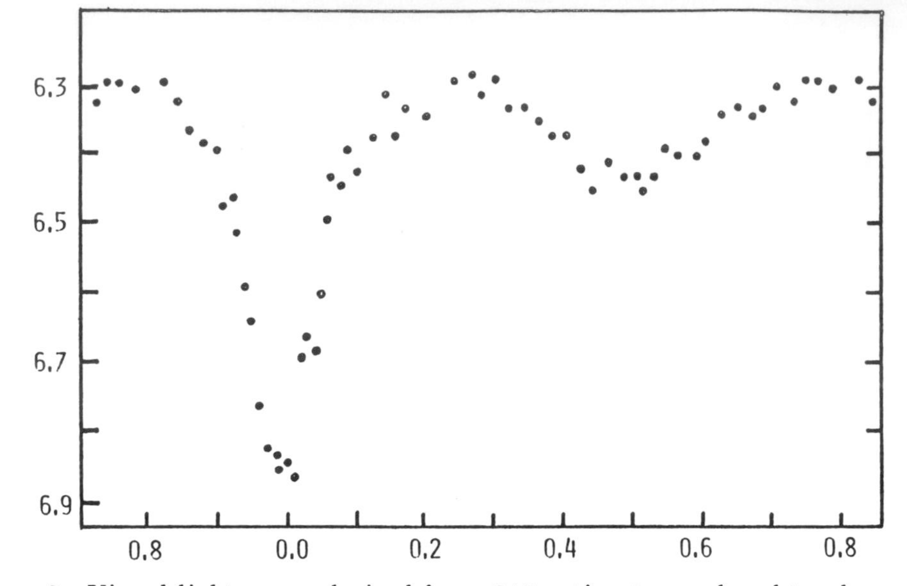

Figure 2. Visual light curve derived from 312 estimates, reduced to phase according to the light elements given in Figure 1. Each point represents the mean phase and magnitude of an average of six estimates in each 0.01 phase interval between phases 0.9 and 0.1, and each 0.02 phase interval between phases 0.1 and 0.9. The light curve's definition is adequate to indicate, with some confidence, that the orbits are circular, the eclipses are total, and the secondary minimum is almost entirely due to ellipticity and reflection effects.

may introduce systematic error. On my light curve, V1010's total range of variation, established from photoelectric studies as $6.^m2 - 6.^m9$, is suppressed by about $0.^m1$ at both maximum and primary minimum. this effect can be attributed to my systematic tendency to overestimate the difference in brightness between two stars when that difference is very small (in this case, when V1010 is only slightly fainter than the $6.^m12$ comparison star at maximum or slightly brighter than the $7.^m16$ star at primary minimum).

One important source of systematic error in visual work, colour perception, is avoided in this case because V1010's colour is virtually constant throughout its cycle of variation. Another source of systematic error, apparent rotation of the instrumental field of view, is minimized by V1010's moderate southern declination, which limits field rotation for an observer at a mid-northern latitude.

(4) It is also helpful to adopt an observing and recording method that permits making distinctions finer than $0.^m1$, the standard interval for most visual

comparison sequences. Because comparison star values to $0.^m01$ were available, I chose to estimate V1010's brightness on an arbitrary scale of ten units between each pair of the three comparison stars, A-B and B-C. Estimates were recorded in the form A-3-v-7-B, B-6-v-4-C, etc., and later reduced to the nearest $0.^m01$.

This decimal interpolation method yields units of just $0.^m04$ when V1010 is brighter than the intermediate comparison star and $0.^m07$ when fainter. The resulting light curve demonstrates that these small units are meaningful in visual work when the conditions outlined above are met.

While pursuing these observations of V1010 Ophiuchi, I also began to collect and analyze previously published times of minima in an effort to clarify difficulties I encountered in defining this star's period prior to 1982. This led to contact with Dr. Edward Guinan, who has been investigating the same question. We have shared our unpublished observations and are preparing an extended period history of V1010 Ophiuchi for publication.

II. PHOTOGRAPHIC OBSERVATIONS

Emilia P. Belserene is Director of the Maria Mitchell Observatory, an observatory which combines the scientific and the educational aspects of astronomy in a unique and effective way. Each summer, the observatory takes on several undergraduate astronomy students, and trains them in the observation and analysis of variable stars. In addition, the students participate in the observatory's active public education program.

In her review, she eloquently explains "what is right about photographic photometry" as well as how to live with what is wrong with it. Even after CCD's (or their successors) become the chosen imaging detector in professional astronomical research, there will still be a million or more archival photographic plates, containing information whose precise nature may not presently be known. In particular, the major sky surveys in both the northern and southern hemispheres are photographic in nature, and may continue to be for many years. Sophisticated scanning devices have been developed to extract information from these photographs, so clearly astronomers must continue to be aware of the potential and the problems of photographic photometry.

Even in the future, photography will continue to be enjoyable and inexpensive, and I am sure that thousands of amateurs and students will continue to indulge in it. Considering the popularity of astrophotography, it is surprising that more people do not use their photographs to advantage — to search for and measure variable stars. There is a need, for instance, to monitor long-period and irregular variables. Those in globular clusters are particularly convenient to study (because several can be recorded on a single image) and easy to interpret (because the distance and other properties of the cluster can be independently determined). Observers with access to a telescope with a suitable plate scale might want to get involved in such a study.

Photography of variable stars in globular clusters has long been a specialty of the David Dunlap Observatory, so it is fitting to have a paper from Christine M. Clement which demonstrates the effectiveness of photographic photometry, and the type of astrophysical data which can be obtained by means of it. The period of a variable star can be determined more precisely than any other stellar property. In the double-mode RR Lyrae stars, two such properties can be determined in a single star!

PHOTOGRAPHY IN A PHOTOELECTRIC ERA

Emilia Pisani Belserene
Maria Mitchell Observatory
3 Vestal Street
Nantucket, MA 02554
U.S.A.

INTRODUCTION

My first taste of photographic photometry came in an undergraduate research project, exploring the possibilities of using a Ross 3 inch camera with a red filter and red sensitive plates, to approximate a photographic red system of magnitudes then in use at Harvard. The project did not succeed in producing good red magnitudes, but I did learn a great deal about why photographic images do not readily tell us how bright the stars are. I resolved not to do photographic photometry ever again.

As it happened, I came back to the method a few years later and have been using it ever since. The problems had not been solved in the meantime. In fact, they seem greater now, by contrast with the kind of work that can be done photoelectrically. Photoelectric photometry has added a wonderful new dimension to variable star research, but it can not do everything. Visual and photographic techniques are still needed.

Janet Mattei (elsewhere in this volume) has reviewed visual methods and programs. I would like to review the place of photography in variable star research: what is wrong with it, what is right with it, how to cope with some of the disadvantages.

What is wrong with the photographic emulsion as a light detector? A very important problem is that there is no simple relationship between the amount of incident light and the effect it causes. There is even the problem of deciding how to measure this effect. Then, there is the graininess, which limits the precision with which the photographic effect can be measured, whatever we decide to use as the measure. There is the low quantum efficiency, that is, the waste of photons, especially when the light levels are low. The optical system contributes problems also: the structure of the image, which influences the deduced magnitude, varies from one part of the plate to another, and is influenced by the colour of the star.

THE ADVANTAGES

There are ways to deal with these problems and I will have more to say about that in a little while, but first let us look at what is *right* about photographic photometry. Why should we bother to use a method with so many problems?

What is right about using a small telescope to study variable stars photographically? Most important, I think, is the number of stars whose images are produced at the same time. Photoelectric work is making some progress in observing more than one star at a time, but photography is many factors of ten better in this regard. A related advantage is the permanence and compactness of the record. A single exposure can produce hundreds of thousands of images, mostly of stars not thought to be interesting at the time the plate was taken. These images form a permanent record of the sky, which we can consult later to learn the past behaviour of objects which have turned out to be interesting. Martha Hazen, curator of the plate collection at Harvard, has called it the Harvard time machine. We can go to the plate stacks and see how the sky looked in 1888, say.

Another great advantage of photography is the option of using long exposure times. This is especially significant at small telescopes, which are not in great demand. They may not have speed but they do have time. With patience it is possible to guide well for as long as needed to produce images of the stars of interest, limited only by skylight or the crowding of the field as more faint stars stand out above the background. With the little 18 cm Cooke at the Maria Mitchell Observatory we regularly study variable stars which go down to 15th magnitude. The plate scale, about 4 arcmin/mm, is too crowded to go fainter in the Milky Way, but at high galactic latitude we can use longer exposures and go down to 16th or 17th magnitude. Some of the most useful of the old plates at Harvard were taken with a 3 inch lens and exposure times of the order of 2 hours.

The small observatory has another motivation to use photography. The technology is relatively inexpensive, not perhaps compared with the eye as a light detector, but certainly compared with a modern rival in two-dimensional imaging, the Charge Coupled Device. It has been said (Furenlid 1984) that "CCD's have superior detective qualities, but are small and expensive, whereas the opposite is true for photographic emulsions." His table comparing the properties of the two imaging methods is reproduced here as Table 1.

A small telescope, then, if it is used photographically, can go deep by using long exposure times. It can record enormous numbers of star images at the same time, and it can store this enormous amount of information compactly and essentially permanently.

Table 1.

Plates	CCD
1. Nonlinear	1. Linear
2. Large size	2. Small size
3. Two-dimensional	3. Two-dimensional
4. Small dynamic range (approximately 1:100)	4. Large dynamic range (approximately 1:10,000)
5. Low sensitivity	5. Extremely high sensitivity
6. Wide colour sensitivity	6. Wide colour sensitivity
7. High density of information; storage in convenient form (images)	7. Information storage problems; destructive read-out
8. Permanent record (plate)	8. No permanent record (tape)
9. Essentially finished product	9. Still under development
10. Low initial cost	10. High initial cost

APPROPRIATE PROJECTS

Given that photography, while an old method, still has some advantages over newer methods of stellar photometry, what sorts of projects lend themselves to it? The Maria Mitchell Observatory looks for projects that meet some or all of these criteria:

1. Since we use student assistents, we want the projects to be of educational benefit. The same motivation affects the choice of student projects at colleges and universitics.

2. Since our small telescope is not subjected to heavy demand for its time we can be willing to work on projects that are not very pressing. We can try to be of service to the astronomical community by filling in gaps in variable star data, to improve statistics for future studies. We need not justify our efforts by any confidence that the results will immediately answer some pressing question.

3. Since our telescope has been in use for 70 years, we favour projects that make use of old plates as well as new.

4. Since our weather is not very good, we like to embark on some projects that do not require new plates at all, but that can make good use of just the old.

Variable stars make excellent student projects, since a student can be put in charge of a whole star, so to speak. Perhaps the information to be gained from that star will be simply an entry in a table of statistics. The star may be just a small part of someone else's research project. But for the student assistant the star is a whole research project. The educational benefit of taking full charge is very great.

Which stars are good to assign in this way? First, suspected variables not yet confirmed. One does hesitate to assign to a student a task that may lead to a negative result, but if the variability is confirmed the satisfaction can be very great, especially if the variability is of an interesting sort. Second, elements of periodic variables that need updating. Like the first, this can be done by observatories without an existing plate collection, taking new plates for the purpose, but often one wants plates taken in other years. Anyone with a project that can be done on a small telescope should keep in mind the great existing collections of plates that may be consulted for information about other years.

In preparing this review I consulted with Martha Hazen, curator of the plate stacks at Harvard. She reminded me that there are about 400,000 plates there. Many were taken in the 1930's and 40's and cover the whole sky frequently to about magnitude 14 or 15. They were taken with small telescopes (the ones I have found most useful were taken with apertures of 3 inches and smaller) and they can form a very valuable adjunct to modern work with small telescopes. There are valuable collections elsewhere, including, of course, the David Dunlap Observatory with its many thousands of plates. There are several European observatories with good collections of photographs for archival work on variable stars.

When elements of periodic variables can be updated very precisely, they provide data of astrophysical importance. Most of the stars I assign to my student assistants are pulsating variables for which we have enough plates to find 6 to 8 figures in the period. We are not just testing published elements or trying to improve linear ephemerides. We are looking to see whether the period has changed. It is possible, with our plates, to detect period changes on the order of a few tenths of a cycle per million years. An observatory without old photographs has the option of doing similar work by appealing to plate archives to get a long enough time base.

A third way to be of use to the astronomical community is to search fields for unknown variables. Here, as in verifying suspected variation, there is the educationally unfortunate possibility of negative results. We have a fine blink comparator, but, even so, I prefer not to ask an assistant to search for new variables until he or she has had time to produce some positive new results on a known variable. On the other hand, the service to the astronomical community is a fine one. Recently an assistant at the Maria Mitchell Observatory turned up a BL Her star in the south galactic pole region (Waugh 1984). This class consists of Population II stars evolving away from the horizontal branch after exhaustion of helium in the core. Their observed period changes will probably provide useful constraints on theories of this rather rapid stage of stellar evolution (Christianson 1983). It was satisfying to add one more to the short list of stars in this short-lived phase of evolution.

What of observatories without a blink comparator? Ben Mayer (1977) has provided an interesting suggestion: a projection blink comparator, ProBliCom, consisting of two Carousel projectors to compare two 35 mm transparencies (colour slides or black and white negatives) of the same field. Anyone can have a blink comparator and contribute to this useful, if often thankless, task.

Are there really many variables, bright enough to photograph with small telescopes, that remain to be discovered? Yes, I am sure of it. Martha Hazen and Dorrit Hoffleit have pointed out to me that the study of variable star fields initiated by Harlow Shapley at Harvard, was discontinued before all of the fields were analyzed. The original motivation had been to measure distances, to refine the scale of the Galaxy. When Shapley realized that interstellar extinction was going to set severe limits on the accuracy with which he could get distances, he turned to more promising avenues of research. Those of us who are interested in the variable stars themselves, and are willing to put in time for no more reward than filling in gaps in variable star statistics, have some work to do. If we choose fields that were in Shapley's surveys (1928) we will find some particularly good early epoch plates at Harvard.

And of course, even in well studied Sagittarius there are new variables all the time: the novae. I am embarrassed to admit that the Maria Mitchell Observatory, which does not blink plates routinely, could have discovered Nova Sgr 1978 and 1984. In both cases we had pre-discovery images.

There is a fourth project I would like to urge on anyone whose small telescope has much free time: the monitoring of stars whose pulsation exhibits more than one mode. The number of observations required to study them adequately is very large indeed. A few of them are bright enough for photoelectric work with small telescopes, and perhaps the information needed to untangle the complex variation should be looked for in data with errors of only 0.01 or 0.02 mag. or so. Nevertheless, high quality photographic monitoring of some of the fainter stars will turn up some interesting information and already has (Clement & Nemec, this volume). The ratios of the periods, when more than one period is present, can be expected to provide valuable constraints on theories of these stars.

I should say, however, that photography should seldom be used for variables with amplitudes under half a magnitude. If the second period has a small amplitude, the effort will be wasted unless the images are very well exposed and are measured with an effective photometer.

A fifth line of research is not suitable for just any small telescope, but only for those of long focal length. Studying variable stars in clusters is a fine use of one of photography's great advantages, the recording of many images at the same time.

It may be appropriate to close my list of stars to work on with some high-amplitude suggestions. Miras are already included, by implication, among periodic variables whose elements may need updating. There are also slow irregular variables that should be monitored. I personally find it difficult to get interested in the irregular variables, but there is always potential excitement. Some of them may turn out to be interacting binaries where mass exchange occurs at variable rates. Many have been listed as irregular more from lack of knowledge than from lack of regularity. Monitoring the irregular variables is a useful service to be performed by small telescopes.

So, let us agree that photography is one of the good ways for small telescopes to observe variable stars, that it has these advantages:

i) the option of long exposure times;

ii) many images at the same time;

iii) permanent record;

iv) moderate technology and cost.

These are useful advantages, but they do not come without a price. Let us look again at the disadvantages and the ways to overcome or live with them.

STRATEGIES

The most basic problem is one of calibration. How do we relate the photographic image to the amount of light that caused it? Photography conspires against us in several ways. There is nothing about a star image that is simply proportional to the amount of light that caused the image. Here, photoelectric photometry has it all over us. Over an enormous range of light levels, the output of a photoelectric device has that most useful property.

Photographers are dealing with a function that is not linear. In fact, there is no straightforward functional form. In general the relation will be different parts of the field, because of effects of the optics, and will be different for stars of different colour. We must give up any hope of comparing our variable star with just one comparison star, as the photoelectric observers could do (because of the absolute linearity of the photoelectric effect) if they could be certain of the comparison star's constancy. It is only because of its possible variability that they need to include a second star, but we, like the visual observers, need comparison stars whose magnitudes span the variable's entire range.

An historical comment might be in order here. An enormous amount of effort has been expended in the past on schemes to calibrate photographic plates or extend magnitude sequences by photographic techniques. We know now that photography can interpolate reliably, but is quite poor at extrapolating to fainter magnitudes. This is not to say that we should never set up a provisional sequence by extrapolation or by photographic methods. The need is so great that an old scheme (Pickering 1891), has been revived and modernized (Racine 1969; Christian & Racine 1983), but the Pickering-Racine wedge is not as good as having photoelectric standards all the way to the faint end. The internal accuracy of the magnitudes may be fine, but there may easily be a systematic error that depends on magnitude. Blanco (1982) shows that the image structure of the primary and secondary images will generally be different. The measuring technique is sensitive to image structure, especially in iris photometry.

What is the best way to establish magnitude sequences for photographic photometry? Stock & Williams (1962) gave the answer in their review in *Astronomical Techniques*: "photoelectrically established sequences should be used whenever possible." One of the nicest things that photoelectric observers can do for photographers is to give us more B and V magnitudes of faint non-variable stars near the variables we are studying. Photographic photometry in a photoelectric era is much better than photographic photometry when it had to stand alone.

A sensible strategy in the photographic study of variable stars is to set up a local sequence in the immediate vicinity of the variable star by whatever means are available, preferably photoelectric. Often there is a photoelectric or other good sequence elsewhere on the plate. Stars of the local sequence can be compared with it once and for all. The star is then compared only with the nearby sequence.

The word "nearby" requires emphasis. The use of photoelectric sequences solves the lack of linearity but that is by no means the whole story. The photographic effect can depend strongly on location of the image on the plate. If we were to limit ourselves only to images close enough to the centre to avoid this field problem, why then we would lose the disadvantage of having a large number of stars on the one plate. Instead we can use a local sequence confidently, as long as we use it consistently, with the assurance that its images are affected by location on the plate in very nearly the same way as the variable's. For some variable star work it really does not matter how well the magnitudes of the sequence stars are known, but for all variable star work it matters a great deal that we use the same set of sequence magnitudes consistently.

This is worth remembering if there is no photoelectric sequence on the same plate as our field. We can still do very fine work on periods of variable stars with a photographically extrapolated sequence or even just an estimated one. We must remember that the internal accuracy of magnitudes like these is much better than the absolute accuracy. We may be able to get excellent periods but not such good amplitudes and even worse mean magnitudes for comparison with other stars such as for distance estimates.

What is the best way to measure a photographic image? I have so far ignored the problem of what we mean by "photographic effect", anyway. How shall we quantify the blackening that is a developed star image? In just what way shall we compare our variable with its sequence, or, in setting up the sequence, how shall we compare its stars with photoelectrically studied stars elsewhere on the plate? Is it best to look at the photographic density? The diameter of the image? Some combination of the two?

The most modern answer, using the highest technology, is to scan the plate with a microdensitometer and analyze the resulting array of densities by computer, *e.g.*, Cudworth's work in M71 (1985). Those of us who are studying variable stars using small telescopes are apt to be technologically limited in other ways as well, so my recommendations come from the past. Without apology, I recommend "eye-balling" it whenever the images to be compared are visible in the same field under low magnification. This method is by far the fastest, of course. Experience with the archival plates at Harvard has led their curator to use it by preference except for modern fine grain plates with symmetrical images having many silver grains per image. In that case she finds a significant improvement in precision with iris photometry. It is interesting, almost amusing, to note that John Graham (1984) found eye estimates just as good as PDS microdensitometry for faint stars in the galaxy NGC300, observed with the 4 m telescope at Cerro Tololo. He suggests that the accuracy is limited by graininess, just as in Martha Hazen's experience with the old plates at Harvard. The error is in the image, not in our way of measuring it.

Other alternatives are density or diameter measurements, flyspankers, and photometers that pass a beam of light through and around the image. To measure just the density or just the diameter has little to recommend it, since focal images display an increase in both quantities with the amount of incident light. The reason for the history of these methods lies in the attempt to have photography stand on its own, before it had photoelectricity as a servant.

As soon as we have recognized that we are going to use the photograph only for interpolation, we want to take both density and diameter into account. We do this automatically in a direct visual comparison when both the sequence and the variable star are in the field of a low power lens at the same time. For comparing images that are not visible simultaneously, the simplest technology is the flyspanker (Stock & Williams 1962). Between it and microdensitometry are the photometers that pass a beam of light through and around the image. The dynamic range is largest when the beam is variable in size (Eichner *et al.* 1947) so variable iris photometers have supplanted those in which the quantity measured is the amount of light subracted by a star image from a beam of constant size (Schilt 1924).

Iris photometry is very good to have available for transferring photoelectric magnitudes on one part of a plate to a set of local sequence stars on another (preferably at the same distance from the centre). At the Maria Mitchell Observatory we still do use flyspankers, and they have the advantage of working well when the images are not symmetrical, but we do sometimes make transfers with iris photometers "off-island", as Nantucketers say, and I have hopes of bringing one to the island in the future.

I have mentioned graininess as a basic limitation in photographic photometry. Is there something we can do about it? The alternative to graininess is lower quantum efficiency, so let us consider these topics together.

The problem is that the star image is made of a limited number of developed silver grains. With a fast emulsion, the number of photons per grain may be acceptably small, but the grains are large: the number of grains per star image is small also, unacceptably so for the faintest stars. It is not always clear that a group of grains is, in fact, a star image rather than a random grouping. It has become increasingly obvious (Hoag *et al.* 1978) that finer grain is more important than faster speed, at least as soon as the background is appreciable. In this case we are not dealing with underexposure. We are not working at the toe of the characteristic curve (log density *vs.* log exposure). We are, however, looking for small *relative* differences: the effect of 11000 photons in the area of a star image as opposed to 10000 in an adjacent area of the same size.

The potential gain in distinguishing marginal images from the background is very great when fine grain plates can be used. For telescopes that can focus in the red there is a very fine grain emulsion, 2415. For photographic refractors there is IIIaJ, which is sensitive about 600 Å further to the red than the completely unsensitized 103aO and IIaO. I have the impression that most photographic astronomy now uses IIIaJ and IIIaF, for their fine grain, rather than emulsions that more closely match the B and V bands.

Fine grain is important but the slower speeds which come with finer grain are difficult for small telescopes. How shall we cope with this problem? Just by long exposures? It is not as simple as that.

The worst of the speed problem is the way that latent images build up. For a given total amount of light, the process works less well if the arrival of the photons is spread over a longer time interval. This is the familiar low-intensity reciprocity failure. It has received important attention in the past 10 - 15 years with the discovery of ways to treat the emulsion. The various hypersensitizing techniques do their work by removing water and oxygen from the emulsion, and, ideally by also incorporating some hydrogen. Most small observatories can manage the technology to do at least some hypersensitizing. Maria Mitchell is just beginning to get into it. We are about to start vacuum drying our plates and keeping them dry during exposure by flowing nitrogen into the telescope tube. The next step would be baking in nitrogen or forming gas (nitrogen with a small amount of hydrogen). We will probably not take this step until we have more space for equipment.

I have touched only tangentially on the wavelength distribution of the light. The solution is not as simple as standardizing the choice of filters and emulsions, as the photoelectric people standardize the choice of their filters and photosensitive surfaces. The most serious problem that I encountered in my undergraduate project was the distribution of wavelengths within the *image*. Stars of similar magnitude but different colours can even interchange their places in the sequence, on plates of longer or shorter exposure. That case was extreme: a photographic refractor was being used outside its wavelength range, and that should be avoided, even for variable star work with a local sequence. Beyond that, the choice of sequence stars not too different in colour from the variable is about all that is required. The universe does not always cooperate with photographic or photoelectric or visual observers in this regard. A century and a quarter ago Maria Mitchell was having difficulty comparing Mira with the bluer star γ Ceti. Her notes say that she decided "to change the star of comparison for one more resembling it in hue" (Belserene 1983). The nearest turned out to be too far away in the sky.

AN EXAMPLE

Let me end by showing light curves that demonstrate both the good and the bad of photographic research on variable stars. Figure 1 contains nine years of data on the RR Lyrae star, EO Com. The scatter is pretty bad. Almost all of it must be in magnitude (± 0.15 or so as expected) rather than in the phases, which have been calculated from a well-determined period. The purpose of the plot was to see how well the time of maximum was predicted by the elements. Obviously there is information on this subject, in spite of the scatter. Figure 2 gives a better idea of just how much information there really is. The same data have been put into twenty overlapping bins and averaged. This is not the place to talk about the various ways of squeezing the greatest possible phase information from such a plot. Suffice it to say that the value of $O - C$ will be known to ± 0.02 cycle or better. With plots

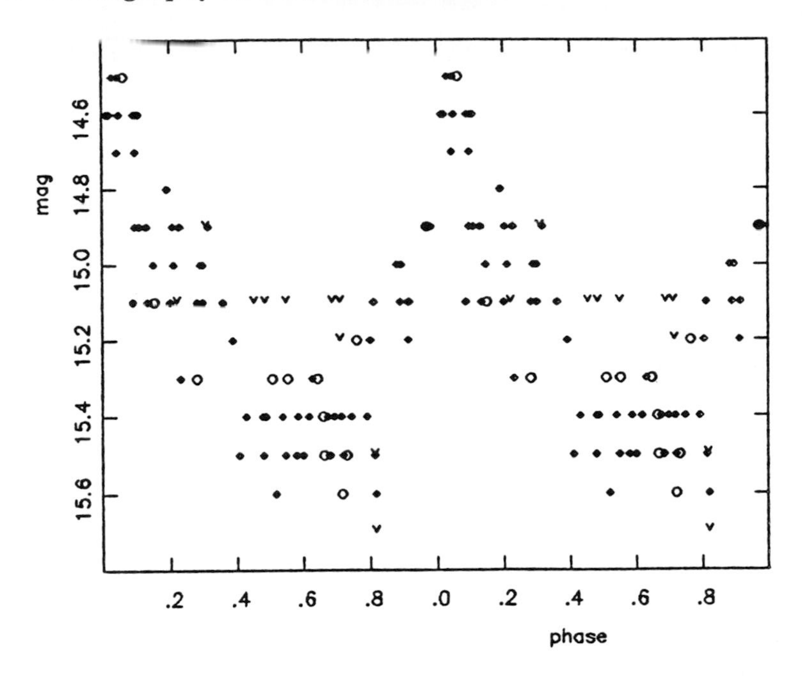

Figure 1. Photographic light curve of EO Com, 1976-1985, according to the heliocentric elements $JD_{max} = 2437352.572 + 0.632090E$, *individual plates. The open symbol signifies lower weight. The "v" indicates a brighter limit.*

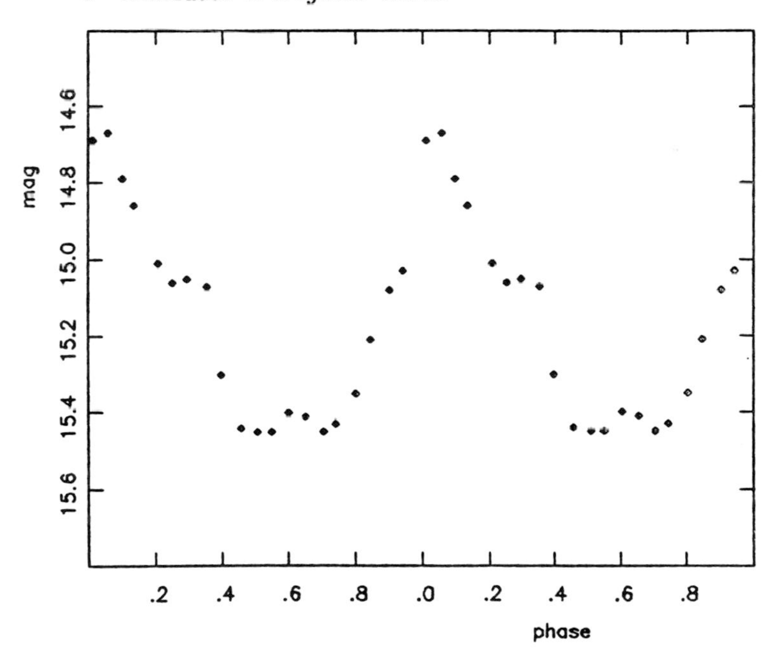

Figure 2. The data of Figure 1 combined into 20 mean points, overlapping phase bins of width 0.1 cycle.

such as this, the student assistants at the Maria Mitchell Observatory are making important contributions to the statistics of the changing periods of variable stars.

Acknowledgements

I am grateful to the National Science Foundation, grant AST 83-20491, for help with variable star work at the Maria Mitchell Observatory. The data in Figures 1 and 2 are from a study by Anthony Reynolds (1985). Conversations with Martha Hazen and Dorrit Hoffleit have been most informative.

REFERENCES

Belserene, E. (1983). *J. Amer. Assoc. Var. Star Obs.*, **12**, 49.
Blanco, V.M. (1982). *Pub. A. S. P.*, **94**, 201.
Christian, C. & Racine, R. (1983). *Pub. A. S. P.*, **95**, 457.
Christianson, J. (1983). *J. Amer. Assoc. Var. Star Obs.*, **12**, 54.
Cudworth, K.M. (1985). *A. J.*, **90**, 65.
Eichner, L.C., Hett, J.H., Schilt, J., Schwarzschild, M., & Sterling, H.T. (1947). *A. J.*, **53**, 25.
Furenlid, I. (1984). *AAS Photo Bulletin*, **36**, 5.
Graham, J.A. (1984). *A. J.*, **89**, 1332.
Hoag, A.A., Furenlid, I., & Schoening, W.E. (1978). *AAS Photo Bulletin*, **19**, 3.
Mayer, Ben. (1977). *Sky and Telescope*, **54**, 246.
Pickering, E.C. (1891). *Harv. Ann.*, **91**, 14.
Racine, R. (1969). *A. J.*, **74**, 1073.
Reynolds, M.A. (1985). *J. Amer. Assoc. Var. Star Obs.*, submitted.
Schilt, J. (1924). *Bull. Astr. Inst. Neth.*, **2**, 135.
Shapley, H. (1928). *Proc. Nat. Acad. Sci.*, **14**, 825 = *Harv. Repr.* **51**.
Stock, J. & Williams, A.D. (1962). In *Astronomical Techniques*, ed. W.A. Hiltner, p. 374. Chicago: University of Chicago Press.
Waugh, J.F. (1984). *I.A.U. Comm., No.* 27; *Inf. Bull. Var. Stars*, No. 2601.

THE DOUBLE MODE RR LYRAE STARS IN THE GLOBULAR CLUSTER IC4499

Christine M. Clement
Department of Astronomy
University of Toronto
Toronto, Ontario
Canada M5S 1A1

James M. Nemec
Department of Astronomy
California Institute of Technology
Pasadena, CA 91125
U.S.A.

The study of double-mode RR Lyrae stars in galactic globular clusters provides valuable information on stellar masses. The simultaneous presence of the first overtone (normally found in the RRc stars) and the fundamental mode (normally found in the RRab stars) has been found in RR Lyrae variables in a few systems. These are the globular clusters Messier 15 (Sandage, Katem, & Sandage 1981; Cox, Hodson & Clancy 1983; Nemec 1985a) and Messier 3 (Goranskij 1981; Cox *et al.* 1983) and the Draco dwarf galaxy (Goranskij 1982; Nemec 1985b). In all known cases, the dominant mode is the first overtone mode. This means that the double-mode pulsators occur among the longest period RRc stars.

The southern globular cluster IC4499 has the highest frequency of RR Lyrae stars of any globular cluster in the Galaxy. These variables were studied by Clement, Dickens, & Bingham (1979). An inspection of their light curves shows a great deal of scatter among the long period RRc stars. Thus the cluster is a prime candidate for an investigation of double-mode pulsation.

Our observational data consist of approximately 100 photographic plates taken at the Las Campanas Observatory of the Carnegie Institution of Washington: 60 with the University of Toronto 0.6 m reflector and 40 with the Swope 1 m reflector. For the investigation, we studied 23 variables. Of these, 20 were known RRc stars and the other three were stars for which periods were not previously determined. The first step in the analysis was to determine the period for the dominant mode (in this case, the first overtone) for each star, using a program of the type described by Stellingwerf (1978). Then the data were "prewhitened". In other words, a mean light curve was fit through the points, and for each point, a residual was computed. The period finding procedure was then applied to the

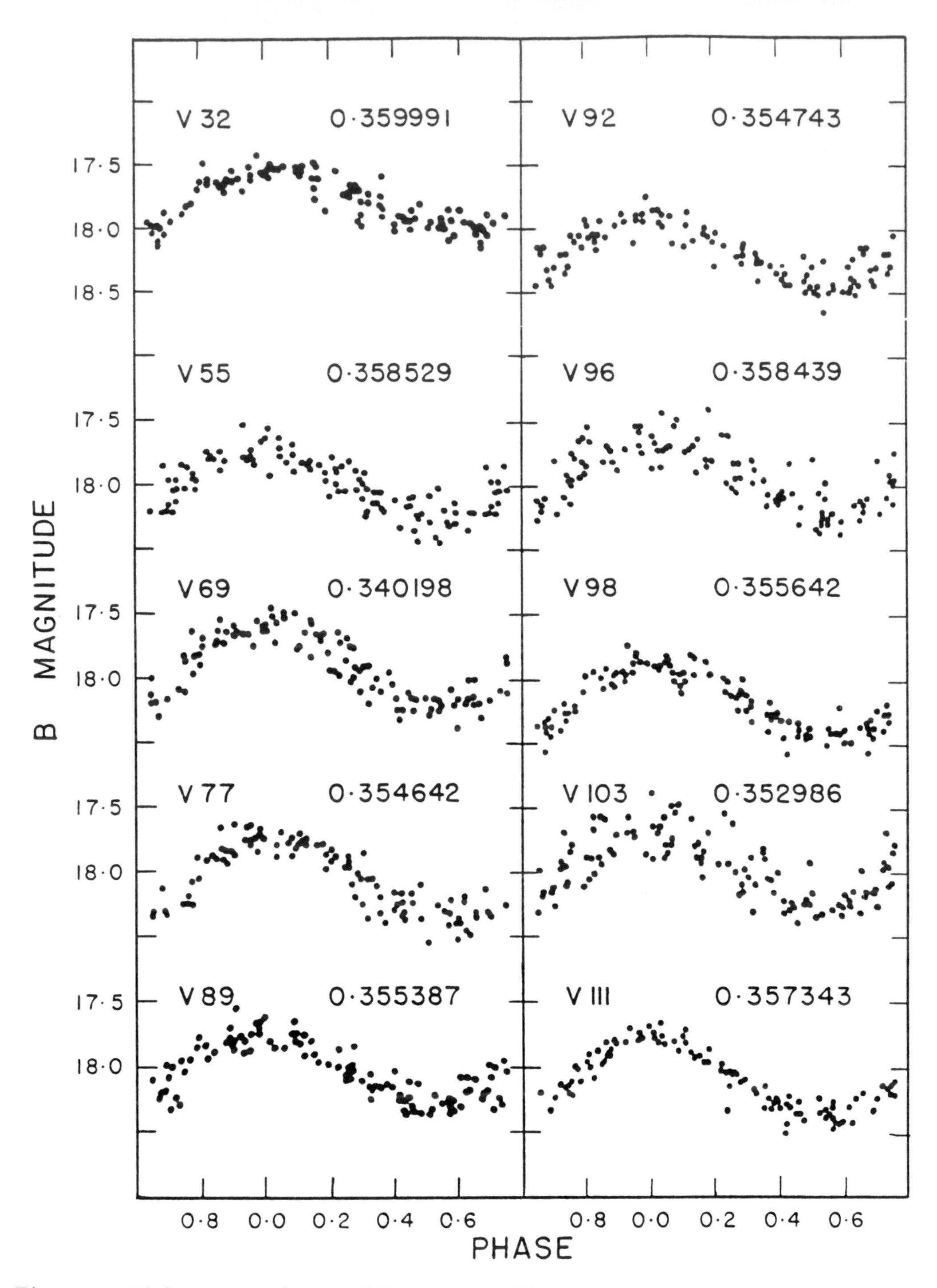

Figure 1. Light curves for 10 RRc stars in IC4499.

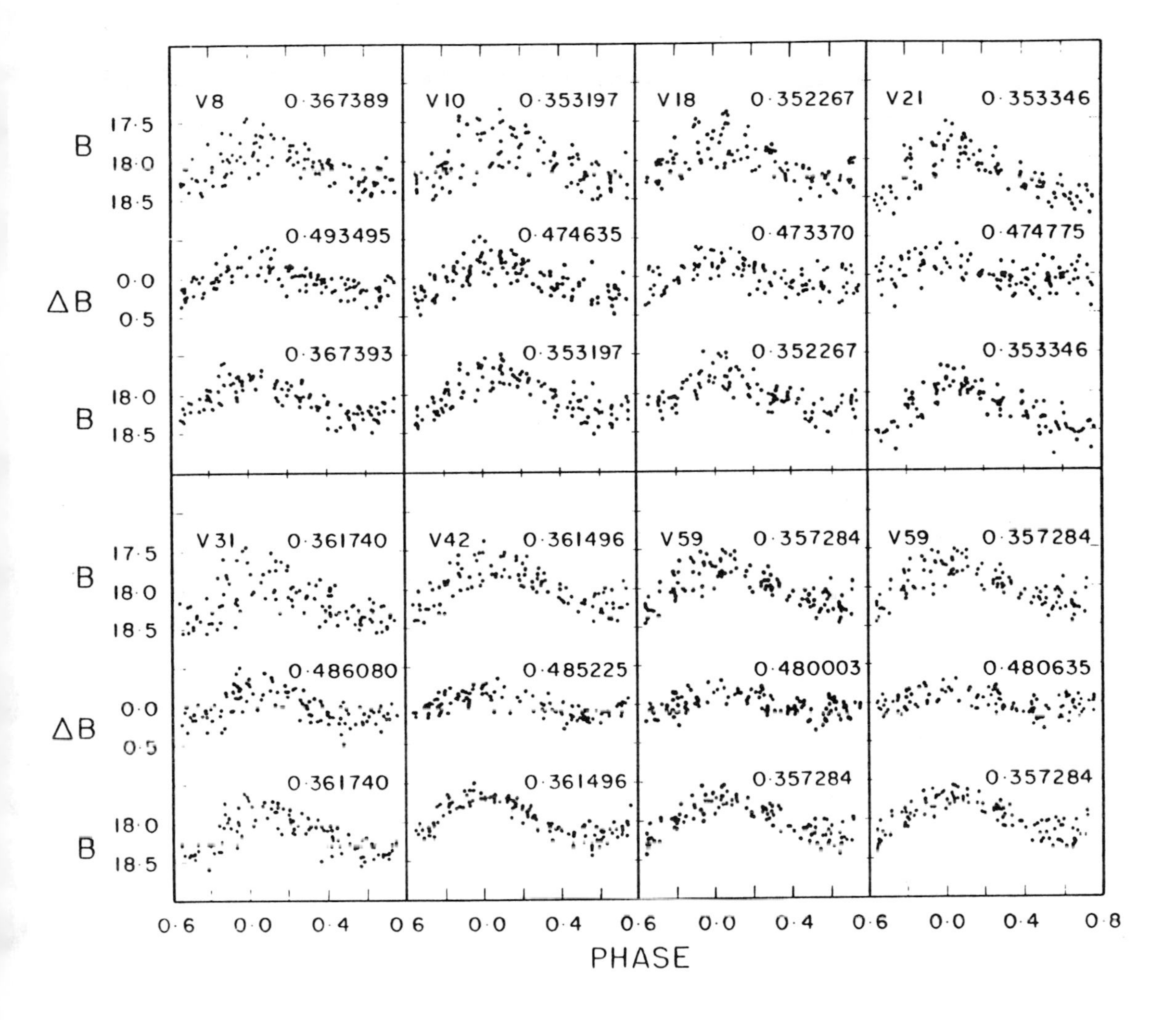

Figure 2. Light curves for 8 of the double-mode RR Lyrae stars in IC4499. The upper curve in each panel shows the original data phased according to the primary (first overtone) period. The middle curve shows the original data prewhitened with the primary period and phased according to the secondary (fundamental) period. The lowest curve in each panel shows the original data prewhitened with the secondary (fundamental) period and plotted with the primary (first overtone) period.

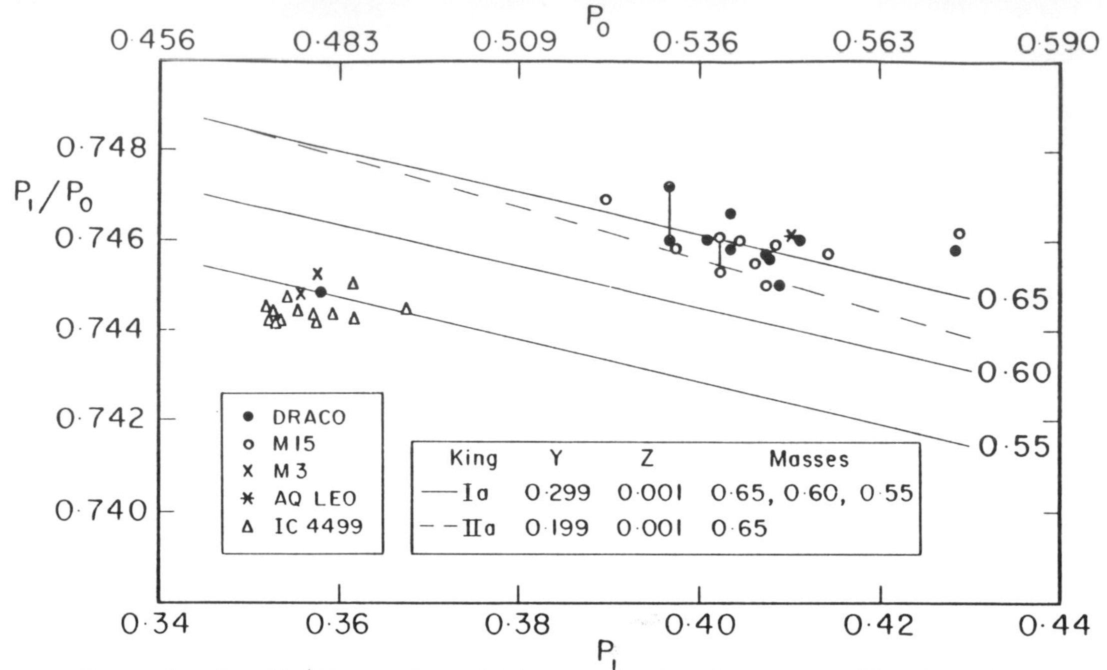

Figure 3. The P_1/P_o vs. P_1 relation, using for the mass calibration the King Ia model used by Cox et al. (1983). P_o and P_1 are the fundamental and first overtone periods, respectively. The double-mode RR Lyrae stars in IC4499, Draco, M15, and M3, and the field star AQ Leo are plotted. The dashed line shows the effect of decreasing the helium content, Y, from 0.3 to 0.2.

residuals to search for periodicity in the range expected for the fundamental mode. The analysis showed that 13 of the variables were double-mode pulsators while the other 10 were pulsating in only one mode. The light curves for the 10 single-mode pulsators are shown in Figure 1 and those for 8 of the double-mode pulsators in Figure 2.

Jorgensen & Petersen (1967) have shown that for radially pulsating stars in which two modes are simultaneously excited, a mass can derived from the observed periods and period ratio. In Figure 3, we show the theoretical relation used by Cox *et al.* (1983) to determine the masses for the double-mode RR Lyrae stars in M15 and M3. We have plotted the double-mode RR Lyrae stars in IC4499, Draco, M15, and M3 on the diagram. It appears that the mean mass for the double-mode RR Lyrae stars in IC4499 is smaller than for those in any of the other systems. Cox *et al.* (1983) have suggested that the mass and metal abundance for RR Lyrae stars in clusters are correlated in the sense that the higher the mass, the lower the metal abundance. The only determination of metal abundance for IC4499 is that of Smith (1984) from spectroscopic observations of a few of the RRab stars. His investigation indicates that the globular cluster IC4499 has a higher abundance than M3 and M15. Our new results for IC4499 therefore give support to the hypothesis of Cox *et al.* (1983). However, the most striking feature of Figure 3 is the way the points are congregated into two separate areas. It would be most interesting to look for double-mode RR Lyrae stars with periods intermediate between these two groups, to see whether such stars exist and to see where they lie on this diagram.

This work will be published in full detail in a paper by Clement, Nemec, Robert, Wells, Dickens, & Bingham in the *Astronomical Journal.*

REFERENCES

Clement, C., Dickens, R.J., & Bingham, E. (1979). *A. J.*, **84**, 217.
Cox, A.N., Hodson, S.W., & Clancy, S.P. (1983). *Ap. J.*, **266**, 94.
Goranskij, V.P. (1981). *Inf. Bull. Var. Stars*, No. 2007.
Goranskij, V.P. (1982). *Astron. Circ.*, No. 1216.
Jorgensen, J.E., & Petersen, J.O. (1967). *Zs. f. Ap.*, **67**, 377.
Nemec, J.M. (1985a). *A. J.*, **90**, 240.
Nemec, J.M. (1985b). *A. J.*, **90**, 204.
Sandage, A., Katem, B., & Sandage, M. (1981). *Ap. J. Suppl.*, **46**, 41.
Smith, H. (1984). *Ap. J.*, **281**, 148.
Stellingwerf, R.F. (1978). *Ap. J.*, **224**, 953.

III. PHOTOELECTRIC OBSERVATIONS

Douglas S. Hall and Edward P. Guinan are two of the most able and active practitioners of photoelectric photometry with small telescopes. Both are experts in the field of binary and variable stars, and both have been instrumental in bringing photoelectric photometry into the realm of the small-telescope observer. Doug Hall is a founder of IAPPP, and both he and Ed Guinan have been faithful supporters of the AAVSO photoelectric program. Their reviews deal with the human as well as the scientific aspects of photometry, and are well worth reading in detail. There would be much to be gained, in fact, by reading them once a year!

Ed Guinan was asked specifically to review the methods of data collection and reduction. By following his instructions, you can obtain data of which you can be confident and proud. Photoelectric photometry is one area of astronomy in which the amateur and the student observer can produce data of "professional" quality — but only if all of the correct procedures are followed.

Mart de Groot has contributed a timely and important paper on P Cygni which, though one of the most interesting and extreme objects in the sky, has been neglected of late by photometrists. Partly as a result of his paper, this star is now well monitored, and the results so far have been interesting and rewarding. Jaymie M. Matthews then describes some of his work on the rapidly-oscillating Ap stars. This new class of variables was discovered using the relatively young technique of high speed photometry. Until recently, this technique was restricted to well-to-do observatories which could afford the necessary data collection equipment. Now, with the advent of inexpensive computers, high speed photometry can be done at many more small observatories.

PHOTOELECTRIC OBSERVING PROGRAMS AND OBSERVATIONS

Douglas S. Hall
Dyer Observatory
Vanderbilt University
Nashville, Tennessee 37235
U.S.A.

I. INTRODUCTION

The title of this symposium is "the study of variable stars using small telescopes" and the title John Percy assigned for my talk is "photoelectric observing programs and observations". Therefore I will talk about using photoelectric photometry to observe variable stars with small telescopes. Photometric techniques other than photoelectric (visual and photographic) are being covered in other talks at this symposium. My emphasis will be on the observations, as opposed to the theoretical interpretation. Although other celestial objects suitable for photometry will be mentioned, variable stars will be the principal ones. And, in my thinking, a 24 inch telescope divides the large from the small.

The first section will be a narrative recounting of my own experience as an astronomer which, as preparation for this symposium made me realize, indeed has been almost entirely one of photoelectric photometry of variable stars with small telescopes. In the course of 23 years since starting graduate work at Indiana University in 1962, I have learned many lessons.

The second part will be a more coherent recapitulation of those lessons. To the extent that they are valuable lessons, I pass them on to you as recommendations you might consider following, to make good use of whatever small telescope you might have at your disposal.

II. MY OWN EXPERIENCE

My introduction to photoelectric photometry was in 1963 at the Agassiz Station of the Harvard College Observatory during a Summer Research Participation Program sponsored by the National Science Foundation. Of the assortment of telescopes there, largely under-utilized throughout the rest of the year, I was assigned to the Clark 24 inch broken-Cassegrain reflector. This nice telescope was "retired from service" two years later and to my knowledge is no longer in existence. It was equipped conventionally for photoelectric photometry (unrefrigerated 1P21 photomultiplier, UBV filters, Brown strip chart recorder) and seemed to work well. Early on, I had a nice leisurely chat with the late Sergei Gaposchkin in his office on the Harvard campus, during which he led me through the (then most recent) 1958 Russian Variable Star Catalogue. With the help of the informative "remarks" section in the back, we immediately saw hundreds of stars, any one of which would make a good project for a 24 inch telescope equipped for photoelectric photometry. Some of the results I got that summer with the 24 inch ended up being published (Hall 1967a, 1968), as did some lunar photometry obtained by George V. Coyne during the bright-moon phases.

All of the data involved in my Ph.D. thesis (Hall 1967b) were obtained with the two 16 inch Cassegrain reflectors at Kitt Peak National Observatory, during two month-long observing runs in 1965. My project required narrow-band photometry of more than a dozen different eclipsing binaries obtained during the hours of totality. Month-long allocations of observing time on any telescope larger than 16 inches would have been out of the question. In later years my graduate students and I used these same two small telescopes on numerous occasions to obtain photometry which resulted in literally dozens of published papers. Abt (1980) has demonstrated that, using the measure scientific papers (published and cited) per dollar (of telescope expense), the two 16 inch telescopes at Kitt Peak were exceedingly more productive than the 36 inch, the 50 inch, the 84 inch, or the 4 meter. Nevertheless, for some reason or reasons not obvious to me, Kitt Peak has decided to retire those two telescopes from service.

After getting my Ph.D. in 1967 and taking a faculty position at Vanderbilt University, my telescope was the 24 inch Seyfert Reflector/Corrector of the Dyer Observatory. A 24 inch aperture is regarded by many nowadays as small. Although this fine instrument is capable of excellent wide-angle photography (suitable for iris photometry, objective prism spectroscopy, and even astrometry), I have used it almost exclusively in its Cassegrain mode, for photoelectric photometry of variable stars. Let me describe three of my observing projects which made good use of it for several years.

The faintest of the four stars which make up the Trapezium in the Great Orion Nebula is the variable star BM Orionis, long known as an eclipsing binary with an orbital period of $6.^{d}47$. The extreme youth of the Trapezium (only $\sim$ 100,000 years) made BM Ori unique as the only known eclipsing binary which could be presumed to be in its pre-main-sequence stage of evolution. I was surprised, therefore, to note that there was no photoelectric light curve in the literature or apparently elsewhere. It was certainly not an inconveniently faint star, being $V = 8.^{m}0$ at maximum and dropping only $0.^{m}6$ at primary minimum. Clearly a small telescope project. The difficulty was the extraneous light of the nebulosity itself, in which BM Ori was imbedded, and the scattered light from its nearest neighbour in the Trapezium, only 11 arcseconds away and 2 magnitudes brighter. Clearly a project for a mechanically and optically superior telescope. The Seyfert 24 inch telescope's rigid mount, steady tracking, good image quality, and high magnification of the photometer's postviewer made it possible to do photometry (on nights of good seeing) with a diaphragm 9 arcseconds in diameter, required to exclude most of the scattered light and minimize the contribution from the background nebulosity. The result was a complete *UBV* light curve, showing both the primary eclipse and the shallower secondary eclipse, described by Hall & Garrison (1969). Subsequent analysis by Hall (1971) showed that the fainter star in the binary was a unique object, highly flattened into a toroidal shape (like a red blood corpuscle) by the differential rotation expected theoretically (Bodenheimer & Ostriker 1970) in pre-main-sequence stars contracting down the Hayashi evolutionary track. Subsequent spectroscopy by Popper & Plavec (1976) and subsequent photometry by Arnold & Hall (1976) confirmed this interesting interpretation.

My work then settled onto the long-period Algol-type eclipsing binaries. These are the fascinating semi-detached systems in which one star (the originally more massive of the two) has expanded as a result of evolution off the main sequence, overflowed its Roche lobe, and begun transferring matter in a gas stream which flows over to the other star in a spiral trajectory and builds up a ring or disk of luminous circumstellar material. Those binaries with relatively long orbital periods, between two weeks and one month, are the ones with the most substantial and prominent disks and hence the most interesting. Those long periods, however, make it very difficult to obtain a light curve with complete phase coverage; the eclipses themselves, from first contact to fourth contact, last well over 24 hours and hence cannot be covered in any one observing night even by observing from dusk to dawn. Typical two-week-long observing runs at Kitt Peak would be woefully inadequate for these stars. We needed persistent night-after-night observing throughout the year and often for more than one year. The 24 inch at Dyer Observatory was ideal for this, and ready accessibility was the key. Dyer is on the south edge of the city of Nashville, with remarkably dark skies to the south, and only a 10 minute drive from my home. The observatory itself has our offices, the machine shop, the electronics shop, and the complete astronomical library. One hardly hesitates to make a 10 minute drive, to take advantage of a 2 hour interval of clear sky. One would, however, hesitate to make an hour-long drive to a remote site unless the sky promised to be cloudless all night long. Statistics have shown (Heiser 1970) that, on an hour-by-hour basis, the sky in this part of Tennessee is totally cloudless about 1/3 of the time, whereas the fraction of nights cloudless the entire night would be much smaller. For almost a decade, Dyer Observatory had an outstanding reputation for its work on these awkward long-period variables. The important eclipsing binary systems for which complete *UBV* photoelectric light curves were obtained include: RS Cephei, SW Cygni, WW Cygni, UZ Cygni, RX Geminorum, RY Geminorum, AQ Pegasi, RW Persei, and BS Scuti.

Most recently my attention has been directed to the fascinating RS Canum Venaticorum-type binaries, variable by a mechanism which even now is not mentioned in most text books. Typically 10% or 20% of the surface of one star in the binary is covered by a huge area of starspots; as the dark region rotates into and out of view, the brightness periodically decreases and increases. These variables make an ideal observing project for a small telescope which is readily accessible and can look at several stars on each night. (1) A large number are quite bright. Among the hundred or so such systems already known, 35 appear in the *Yale Catalogue of Bright Stars.* (2) Because their periods are days or weeks in length and their light curves are smooth, only one photometric observation per night is required. (3) The amplitude of the variability is between $0.^m03$ and $0.^m3$, tractable for photoelectric photometry but impossible for visual or photographic photometry. (4) The variability is sufficiently periodic that several incompletely observed cycles can be pieced together to achieve complete phase coverage. (5) The light curves do, however, change gradually from year to year (in a manner not yet understood) and therefore cannot be considered to have been defined once and for all, no matter how good the photometry. Thus, the same favourite stars remain as objects of interest year after year. (6) There are two or three different reasons for suspecting that supercycles, a decade or so in duration, control the evolving shape of the light curve. The goal of defining these elusive supercycles means that persistent observing programs lasting many years are likely to be rewarded.

In 1968 I discovered my first amateur photoelectric photometrist. It was Larry P. Lovell of Chagrin Falls, Ohio, working with a 10 inch Cassegrain telescope in his backyard observatory which he called the Hickox Observatory. The electronic equipment and the photometer itself had been provided by his neighbour Arthur J. Stokes in nearby Hudson, Ohio. From previous work he had done, it was apparent to me that an amateur could have photoelectric equipment comparable in sophistication and resultant photometry comparable in accuracy to that commonly associated with professional astronomers. It seemed to me that such an observatory lacked only direction, *i.e.*, what stars to observe for maximum benefit. Lovell and I thereupon began a collaboration whose progress can be traced from the first paper we co-authored (Lovell & Hall 1970) to the most recent (Scaltriti *et al.* 1984), a total of 22 to date. Obviously an observatory like Hickox will be limited to somewhat bright stars because of the limited light gathering power of the 10 inch mirror, the relatively bright skies at the backyard site, and the dark current from the uncooled photocell. But both of us were surprised to realize how many bright (in fact, *very* bright) stars needed (and still need) photoelectric photometry. Our 1970 paper presented the first photoelectric photometry of β Lyrae obtained anywhere in the world since 1959.

In June 1980, along with Russell M. Genet, I founded I.A.P.P.P. after the number of my small-telescope collaborators had grown to a couple of dozen and an even larger number of potential amateur and small-college photoelectric observers were out there waiting to become mobilized. The initials stand for International Amateur-Professional Photoelectric Photometry. The organization publishes the quarterly *I.A.P.P.P. Communications.* The inside front cover of each issue explains: "I.A.P.P.P. was formed in June 1980 to foster communication on various practical aspects of astronomical photoelectric photometry. The role of IAPPP is to supplement and support the work of other organizations and established journals, by concentrating on the practical details (equipment, observing techniques, data reduction, computer software, and observing programs) not normally discussed at symposia or published in other journals. IAPPP does not publish observational data as such, because the ultimate goal of IAPPP is to see reliable and scientifically useful photoelectric photometry from smaller observatories appear in reputable astronomical publications where it will be accessible to the astronomical community." It was intended that professional astronomers would describe scientifically worthwhile observing projects suitable for small telescopes — not merely suitable but in many cases able to be done *best* by a small-telescope backyard observatory or by a coordinated network of such observatories. Amateurs and others with small telescopes would read the descriptions in current issues of the *I.A.P.P.P. Communications*, select one or more that seemed suitable, collaborate directly with the professional, and co-author papers describing the results in a professional astronomical publication. As one such professional, I myself have continued my collaborations with amateurs. There is no one centralized I.A.P.P.P. photoelectric observing program. Its activity is the sum total of the individual and collaborative efforts of its amateur and professional members. Unexpected but exceedingly beneficial was the reverse interaction. Amateur astronomers who are professionals (or retired professionals) in other fields — such as optics, electronics, instrumentation, or computer systems — began making advances in photoelectric photometry techniques which put them ahead of what was being done by professional astronomers. These advances, described also in the *I.A.P.P.P. Communications*, are now being assimilated by the astronomical community at large.

As of the writing of this paper (September 1985) I.A.P.P.P. has 645 members, about evenly divided between professionals and amateurs, in 45 different countries. Twenty-one issues of the *I.A.P.P.P. Communications*, published quarterly, have been distributed and the next issue is in preparation. They are edited at Dyer Observatory, printed by The Master Printer in Nashville, and mailed from Vanderbilt University. A very large number of observing projects, listed in Table 1, have been described in the *Communications.* Most, though not all, involve stars and most, though not all, are suitable for telescopes of quite small aperture. One measure of how successful small telescopes have been at accomplishing scientifically worthwhile observing projects is the count, in Table 2, of papers based on photoelectric photometry and co-authored by amateur astronomers. Needless to say, this list, which does not include papers which have only professionals as authors, is a small subset of all papers based on photoelectric photometry with small telescopes. I.A.P.P.P. Symposia and Workshops, several each year, have been held since the first in Dayton, Ohio in June 1980. Countries included have been Canada, England, France, New Zealand, the U.S., and Yugoslavia. They have been held often in conjunction with other organizations, both local and national; those of the latter including the Americal Astronomical Society, the Astronomical Society of the Pacific, the Astronomical League, the Royal Astronomical Society of New Zealand, and (most recently, here in Toronto) the Royal Astronomical Society of Canada and the A.A.V.S.O. The last one was in Flagstaff, Arizona and the next one will be in Melbourne, Florida in December 1985.

For my last experience to include in this paper I mention the so-called Automatic Photoelectric Telescope, designed and built by Louis J. Boyd with subsequent collaborative assistance from Russ Genet. This humble, home-built 10 inch Newtonian in Louis' backyard in downtown Phoenix works automatically under computer control from sunset to sunrise, deciding which stars to observe and in which sequence, finding the stars, centering them in the photometer diaphragm, timing integrations, switching filters, and recording all pertinent data on diskette. Recent descriptions of its operation are in Boyd, Genet, & Hall (1984, 1985) and Boyd (1985). With my own involvement in it being that of scientific advisor, we have published 20 papers after less than two years of telescope operation. The essential ingredients of I.A.P.P.P. are the same ones responsible for this significant technical accomplishment: photoelectric photometry, a small telescope, a talented amateur, and interaction with professional astronomers. Although Russ Genet, in a later paper at this same Symposium, will talk in more detail about the Automatic Photoelectric Telescope, let me conclude by saying that the potential for gathering data, which the telescope has demonstrated already, has made it possible to consider ambitious (but needed) photometric observing projects which heretofore could not reasonably have been contemplated. Among those contemplated are (1) monitoring year by year the migrating wave in most of the known RS CVn variables, not just a few selected ones, with one telescope in the *UBVRI* bandpasses, (2) discovering new variables wholesale among a relatively large list ($\sim$ 100) of suspects, (3) monitoring the photometric behaviour of the extremely bright ($V < 3.^m0$) stars which up until now have been neglected because they are *too* bright, (4) determination of *uvby* magnitudes of all the stars in the *Yale Catalogue of Bright Stars*, (5) photoelectric light curves of Mira variables in the *UBVRI* bandpasses, and (6) determination of *UBV* magnitudes of all the stars in the *Henry Draper Catalogue*, although this would require more than one automatic telescope to do the job.

III. LESSONS LEARNED

1. A telescope not used is a telescope wasted. Just as there are more small telescopes in the world altogether, it is not surprising that there are more unused, hence wasted small telescopes. Turning the statement around, there is more untapped potential among the small telescopes of the world than among the large.

2. Convenient access to a telescope makes for more use. Prime examples of conveniently accessible telescopes are telescopes literally in the back yard, telescopes on a college campus proper, and portable telescopes designed to be set up and dismantled quickly. The best dark sites necessarily are usually relatively inaccessible and, if never or rarely used, for that reason are not the best choice.

3. When working on sufficiently bright stars, small telescopes are capable of the same photometric accuracy as large telescopes. The realistic limit for differential photometry, under conditions of good sky transparency, seems to be approximately $\pm 0.^m003$ (see Landis, Louth, & Hall (1985)). Other things being equal, bright stars are inherently more useful to study because for them the additional data needed for a full understanding of the physical processes operating in them (high dispersion spectroscopy, radio, far ultraviolet, x-ray, polarimetry, astrometry, interferometry, etc.) are likely to be available. One should not forget that there almost 10,000 stars in the *Yale Catalogue of Bright Stars* ($V < 6.^m5$) and over 200,000 stars in the *Henry Draper Catalogue* ($V < 9.^m0$) not counting the *Henry Draper Extension.*

4. Due to photocell saturation problems and coincidence problems with pulse-counting equipment, stars can be too bright for a large telescope (although, it must be admitted, a mask could be used to stop down the Palomar 200 inch and make its effective aperture that of a 2 inch telescope, so that photometry of Sirius could be carried out). To this day the brightest 100 or so stars are relatively poorly-studied photometrically for this reason.

5. If light gathering power is not important for a given photometric problem but some other telescope property or feature is, these in general will easier and/or cheaper to achieve with a small telescope than with a large telescope. Examples mentioned in the previous section are the Dyer Observatory 24 inch telescope and Boyd's automatic 10 inch telescope.

6. The recent rapidly increasing sophistication of the optical and electronic equipment which goes on the end of a telescope has made it possible for smaller telescopes to do what larger telescopes could do in the past. I am

told that, in high dispersion spectroscopy of faint sources like quasars, the 48 inch at the Dominion Astrophysical Observatory regularly outperforms what the Palomar 200 inch did only ten years ago.

7. For those wanting to do photometry of variable stars with small telescopes, the prospect is far better than only a few years ago, with respect to learning how to do it without prior expertise or experience and then doing it at reasonable cost and effort. First, there are now a number of good books available, either specifically or closely related. These are listed in Table 3. Second, the required equipment available (photometer heads, photodetectors, electronics, recording devices, computers) is much better, much cheaper, and much more convenient to obtain than before. Photometers, and in some cases complete photometric systems which include everything downstream of the telescope needed for photometry, are available commercially. Examples of the firms are Optec, EMI Gencom, Nielsen-Kellerman Company, Hopkins-Phoenix Observatory, and Pacific Precision Instruments. One of these, a complete photometric system including digital output, costs only about $750. Third, the techniques required to make photometric observations and the techniques required to reduce the data are explained in the same books listed in Table 3. The information contained in those books is, in a manner of speaking, continually updated by means of articles on photoelectric equipment, techniques, etc. which appear quarterly in the *I.A.P.P.P. Communications.*

8. Long-term projects are especially good for small telescopes. This is because of the convenient access characteristic of many small telescopes and the likelihood that they can be dedicated to some one long-term photometric project. It makes sense to use your telescope to do what cannot be done at national facilities like Kitt Peak (because of the typically short allocations of telescope time) or at university observatories (where there is often fierce competition among the faculty researchers and their students for available observing time.)

   Long-term projects often profit from the concept of the cooperative campaign: several different telescopes observe the same variable and pool their results, with nights unobserved at one often observed at another, making for phase coverage as complete as possible. To the extent that there are more small telescopes than large telescopes, wide participation can be expected in cooperative observing campaigns on small-telescope (bright-star) projects. An excellent example is the recent ϵ Aurigae 1982-1984 Eclipse Campaign (Hopkins 1984), in which the photometric coverage proved to be almost entirely a product of small telescope (most of them amateur) observations, because ϵ Aurigae is so bright, $V = 4.^m0$.

Any photometric project involving truly irregular and/or unpredictable variables requires necessarily an enormous amount of monitoring, because in a sense the job is never finished. Few professional observatories, especially those with large telescopes and those where observing time is split between photometry and other tasks like spectroscopy or photography, are willing to devote that many hours and nights of observing time, even though the science behind these irregular variables may be just as important as that behind variables which happen to be periodic. Therefore photometric projects focussed on irregular variables, if they happen to be bright enough, will be naturals for observatories with small telescopes.

9. The geographical location (terrestrial latitude and longitude) of an observatory can make it critically important, independent of telescope aperture.

   If a variable happens to have a period which is an integral number of days, then photometry from any one observatory (at a particular longitude) will produce a light curve with gaps in the phase coverage, corresponding to the daylight hours, which are unavoidable even though the observations are made from dusk to dawn throughout an entire observing season. The remedy is a campaign involving observatories at several points on the Earth's surface, with complementary longitudes.

   If one wishes to obtain a complete light curve of an important non-repetitive variable over some one cycle of its variability, without having to piece together fragments of different cycles, again the solution is a campaign, involving observatories all over the world with a good distribution in longitude. An excellent example was the 1959 campaign on the uniquely peculiar eclipsing binary β Lyrae (Larsson-Leander 1969, 1970).

   Even if a certain celestial object needs photometry only once per night, perhaps it cannot tolerate a gap of one or more nights due to cloud cover, or electronic or mechanical malfunction, or inability of the observer to work on a particular night. Examples would be a unique non-repetitive event like a nova outburst or ground-based back-up of an object simultaneously under observation by a radio telescope, a satellite observatory, or a very long baseline interferometer.

   Finally, where tens of meters on the Earth's surface can make a critical difference, we have the variety of occultation events when a fairly bright star is occulted by the moon's limb, a planet, or an asteriod and when the satellites of Jupiter eclipse and occult each other.

10. It is important, in my opinion and probably most will agree, that your small telescope be used on a scientifically worthwhile problem. Otherwise it might

be said that you are "just playing", or "just collecting more seashells" or "missing the target" or "wasting your shots". In addition to being important, it can be more satisfying in a deeper sense. My experience tells me it is unwise to carve an observing project in stone. This is because the precise direction of scientific interest and relevance is changing rapidly and constantly, and one must be able to follow where it goes. So, how to select and formulate a scientifically worthwhile project which is good for you? There probably is no one best answer.

Surely one good practice would be to subscribe to the *I.A.P.P.P. Communications* and read the various articles appearing quarterly which describe photometry projects suitable for small telescopes. For example, look again at the list in Table 1. For various reasons some of the projects will be better or more appropriate than others, but excellent ones will be there.

Various national organizations of amateur variable star observers, though originally involved with visual estimates of brightness, now have stars on their observing programs specifically intended for photoelectric photometry. Examples are the A.A.V.S.O. in the U.S., the B.B.S.A.G. in Switzerland, the B.A.V. in West Germany, and the A.F.O.E.V. in France. Here one should note whether or not the observing programs are current or updated periodically and whether some stars have been placed on the program as practice objects.

If you attend annual meetings of astronomical societies like the American Astronomical Society, you will hear talks by astronomers presenting their most current scientific investigations. At those talks, you might pick up hot tips which would make for an appropriate and timely photometric observing project.

Regular reading of the scientific literature often reveals, typically in the concluding paragraphs of a paper, a specific note to the effect that photoelectric photometry is required to resolve some question raised but not answered in that paper. Don't assume automatically that someone else will do the job, although you probably should contact the author directly to see if he knows such work is underway elsewhere, in which case you may want to coordinate a joint effort.

Reading books on the general topic of variable stars can help you form an impression of what the major, to some extent long-standing problems in variable star research are. Although such an approach goes against my "carved in stone" warning, there are some photometric observing projects which almost surely will continue to be relevant for years to come. One example is photoelectric times of minimum light for eclipsing binaries with variable (but still not understood) orbital periods.

If there happens to be an astronomy department at the college or university nearest where you live, it might happen that a valuable one-on-one collaboration can be arranged: you supply the photometric data and the astronomer there provides the scientific direction and subsequent analysis. That individual's interests may not be appropriate for a small telescope photoelectric project and a collaboration may not work out but, if it does, the benefits of the close contact and direct communication can make it virtually ideal.

Sometimes the informal exchange of information, which occurs continuously among the 600+ members of I.A.P.P.P., can result in an idea or tip by one person being passed on directly to another for execution. A recent example of this was the tip (by Francis C. Fekel at Goddard Space Flight Center) that the long-period spectroscopic binary τ Persei might undergo eclipse during its upcoming periastron passage. With a lead time of only a few days, he and I were able (by means of the CompuServe network and long distance telephone) to alert a handful of appropriate small telescope observatories which did observe and detect the predicted eclipse.

IV. CONCLUSION

Let me conclude with the statement which I (and others) have made before but which remains true: there are more stars than astronomers. Specifically, there are more relatively bright variable stars requiring photoelectric photometry by small telescopes than there are active observers with small telescopes equipped for photoelectric photometry — far more.

REFERENCES

Abt, H.A. (1980). *Pub. A. S. P.*, **92**, 249.
Arnold, C.N. & Hall, D.S. (1976). *Acta Astr.*, **26**, 91.
Bodenheimer, P. & Ostriker, J.P. (1970). *Ap. J.*, **161**, 1101.
Boyd, L.J. (1985). *Byte*, **10**(7), 227.
Boyd, L.J., Genet, R.M., & Hall, D.S. (1984). *IAPPP Comm.*, No. 15, 20.
Boyd, L.J., Genet, R.M., & Hall, D.S. (1985). *Sky and Telescope*, **70**, 16.
Hall, D.S. (1967a). *Pub. A. S. P.*, **79**, 630.
Hall, D.S. (1967b). *A. J.*, **72**, 301.
Hall, D.S. (1968). *Inf. Bull. Var. Stars*, No. 281.

Hall, D.S. (1971). In *New Directions and New Frontiers in Variable Star Research*, IAU Colloquium No. 15, p. 217. Veröff. der Remeis-Sternwarte Bamberg **9**, No. 100.
Hall, D.S. & Garrison, L.M. (1969). *Pub. A. S. P.*, **81**, 771.
Heiser, A.N. (1970). *J. Tennessee Acad. Sci.*, **45**, 19.
Hopkins, J.L. (1984). *IAPPP Comm.*, No. 17, 22.
Landis, H.J., Louth, H., & Hall, D.S. (1985). *Inf. Bull. Var. Stars*, No. 2662.
Larsson-Leander, G. (1969). In *Non-Periodic Phenomena in Variable Stars*, ed. L. Detre, p. 443. Budapest: Academic Press.
Larsson-Leander, G. (1970). *Arkiv for Astronomii*, **5**, 253.
Lovell, L.P. & Hall, D.S. (1970). *Pub. A. S. P.*, **82**, 345.
Popper, D.M. & Plavec, M. (1976). *Ap. J.*, **205**, 462.
Scaltriti, F. *et al.* (1984). *Astr. Ap.*, **139**, 25.

Table 1. Observing Projects in the *I.A.P.P.P. Communications*.

The name of the project is preceded by the issue number and followed by the name (in brackets) of the proposer.

3. Help Urgently Needed with the RS Canum Venaticorum Binaries. (Hall)
3. Wanted: UBV Photometry of M31. (Keel).
4. Neglected Pulsators: The RV Tauri Stars. (Dawson).
4. International Program for Photometry of Bright Be Stars. (Harmanec, Horn, & Koubsky).
5. Observing Project: U Cephei. (Hall & Olson).
5. Variable Asteroids. (Binzel).
6. A Request for Photoelectric Observations of Cepheid Variables. (Diethelm).
6. Why not Observe a Mira Variable or Two? (Marino).
6. UX Comae: A Wild Eclipsing Binary. (Popper).
7. A Campaign to Observe Eclipsing Binaries with Long Periods. (Diethelm).
7. A FORTRAN Subroutine for Determining Times of Minimum Light. (Mallama).
8. The Need for Cometary Photometry. (Marcus).
9. Comet Photometry and the International Halley Watch. (Edberg).
9. The ϵ Aurigae Eclipse. (Hopkins).

9. New Variable Stars Discovered Photoelectrically by Amateurs. (Hall).

10. Reduction of Photoelectric Lunar Occultation Data. (Blow).

10. Fight Light Pollution with Your Photometer. (Dawson).

10. Prediction of Minimum of Eclipsing Variables — A New BASIC Program for Small Computers. (Hasler).

11. Photometry of T Tauri Variables at Van Vleck Observatory. (Herbst).

11. UU Cancri — The Twin of 5 Ceti. (Kalv).

11. Appeal for Observations of RZ Draconis. (Kennedy).

11. Photoelectric Challenges among the Asteroids: 8 Flora. (Pilcher).

13. Opportunities for Photoelectric Photometry of Asteriods During 1983 and 1984. (Harris).

13. Photometry of Asteroids. I. (Cunningham).

13. Twenty Bright Suspected Variables. (Hall).

13. RR Tel: An Interesting Southern Variable. (Heck & Manfroid).

13. Atmospheric Transparency and the El Chichon Volcanic Eruption. (Miles).

14. Important Miras for Small Telescopes in the Northern Hemisphere. (Wing & Hall).

14. How Can We Learn More About the Dark Starspots in RS CVn Binaries. (Eaton).

14. An Invitation to Lunar Photometry. (Westfall).

14. The Occultation of 1 Vulpeculae by the Minor Planet Pallas. (Douglas, Parker, Beish, Martin, & Monger).

14. Asteroid Photometry with the DOAA SSP Photometer. (Schmidt).

15. Photometry of Asteroids. II. (Cunningham).

16. The W Serpentis Stars. (Wilson, Rafert, & Markworth).

16. The V1010 Ophiuchi Binaries. (Shaw).

16. RZ Eridani Needs Your Help. (Hall).

16. On the Non-Variability of Capella. (Krisciunas).

16. 8 Flora — A Minor Planet Observing Project. (Hollis).

17. Update on ϵ Aurigae (May 1984). (Hopkins).

17. The 1983-84 Halley's Comet Observing Season and the Comet Crommelin Trail Run. (Edberg).

17. Near-Infrared and Optical Light Variations of Symbiotic Stars. (Slovak).

17. The Photometry of Asteroids. III. (Cunningham).

19. Small Amplitude Red Variables: A Brief Introduction. (Percy).

19. Occultations of the Galilean Satellites in 1985 and 1986. (Asknes & Franklin).

19. Photometry Needed for the New ς Aurigae System 22 Vul. (Parsons & Ake).

20. Catch the Minima: A Real Challenge on V348 Sgr. (Heck & Manfroid).

20. Forty Suspected Variable Stars. (Fekel & Hall).

21. Suspected Variability of 10 (θ) Piscium. (Pazzi).

21. A Chromospherically Active Star Suspected of Variability. (Bidelman).

21. Help! Photometry of T Tauri Stars Needed. (Herbst).

Table 2. Papers Published by Amateurs.

Journal	Papers
Acta Astronomica	4
The Astronomer	1
Astronomical Journal	5
Astronomy and Astrophysics	2
Astrophysical Journal	2
Astronomical Journal Letters	2
Astrophysics and Space Science	8
B.A.V. Rundbrief	11
Bulletin of the American Astronomical Society	7
Byte	1
Information Bulletin on Variable Stars	64
Journal of the American Association of Variable Star Observers	23
Journal of the Royal Astronomical Society of Canada	1
Journal of the Royal Astronomical Society of New Zealand	1
Memorie della Societa Astronomica Italiana	1
Minor Planet Bulletin	7
Publications of the Astronomical Society of the Pacific	7
Sky and Telescope	1
Total	148

Table 3. Books on Photometry with Small Telescopes.

Photoelectric Astronomy for Amateurs
F.B. Wood
(1963). New York: The Macmillan Company

Manual for Astronomical Photoelectric Photometry
A.J. Stokes, D. Engelkemeir, L, Kalish, J.J. Ruiz
(1967). Cambridge: A.A.V.S.O.

Photoelectric Photometry of Variable Stars
D.S. Hall & R.M. Genet
(1982). Fairborn: I.A.P.P.P.

Astronomical Photometry
A.A. Henden & R.H. Kaitchuck
(1982). New York: Van Nostrand Reinhold

Software for Photoelectric Photometry
S. Ghedini
(1982). Richmond: Willmann-Bell

Advances in Photoelectric Photometry, Volume I
R.C. Wolpert & R.M. Genet
(1983). Fairborn: Fairborn Observatory

Micro Computers in Astronomy, Volume I
R.M. Genet
(1983). Fairborn: Fairborn Observatory

Solar System Photoelectric Photometry
R.M. Genet
(1984). Richmond: Willmann-Bell

Advances in Photoelectric Photometry, Volume II
R.C. Wolpert & R.M. Genet
(1984). Fairborn: Fairborn Observatory

Micro Computers in Astronomy, Volume II
R.M. Genet & K.A. Genet
(1984). Fairborn: Fairborn Observatory

Lunar Photoelectric Photometry Handbook
J.E. Westfall
(1984). San Francisco: A.L.P.O.

Micro Computer Control of Telescopes
R.M. Genet & M. Trueblood
(1985). Richmond: Willmann-Bell

ACQUISITION, REDUCTION AND STANDARDIZATION OF PHOTOELECTRIC OBSERVATIONS

Edward F. Guinan, George P. McCook, & Joseph P. McMullin
Department of Astronomy
Villanova University
Villanova, PA 19085
U.S.A.

I. INTRODUCTION

Astronomical photoelectric photometry is one of the few areas in modern science where amateurs, students, and small college observatories can participate in research and make valuable contributions to science. This is demonstrated by the number and quality of papers on photoelectric photometry done with small telescopes reported in the professional literature. This photometry has led to a better understanding of the light variations of pulsating stars, B-emission stars, the determinations of the orbital and physical properties of eclipsing binaries, and the discovery of huge starspots and spot cycles on the surfaces of rapidly rotating components of RS Canum Venaticorum and BY Draconis variables, just to name a few examples.

Technological advances in solid state electronics, microprocessors, photoelectric detectors as well as in telescope design and construction have made it possible to obtain research quality photoelectric photometry systems with a reasonable expenditure of time and money. Over the past 5 years, the availability of commercial "off the shelf", light-weight photoelectric photometers at modest cost ($700 - $2000) has made photoelectric photometry accessible and affordable for many who have small telescopes but who don't have the time or the experience to build their own equipment. Because of this, the number of amateur and student astronomers who are actively engaged in photoelectric photometry has increased from about a dozen people about 10 years ago to nearly a hundred now. Moreover, it appears that the overall quality and the scientific usefulness of the photometric observations have also increased. The higher precision of modern photoelectric photometry is due, in part, to improvement in the data acquisition and recording equipment but also arises from a more scientifically and technologically informed observer.

The increasing interest and activity in photoelectric photometry can be gauged roughly by the growth in organizations such as the A.A.V.S.O (PEP section) and the I.A.P.P.P., as well as by the success of a few journal devoted entirely to photoelectric photometry — the *I.A.P.P.P. Communications* — and the success of several new books — *Photoelectric Photometry of Variable Stars*, by D.S. Hall

& R.M. Genet, *Astronomical Photometry*, by A.A. Henden & R.H. Kaitchuck, *Advances in Photoelectric Photometry*, by R.C. Wolpert & R.M. Genet and several others.

In this paper, we will discuss observational and reduction techniques that are important in achieving accurate and useful photoelectric photometry. To expedite the data reduction and standardization, a relatively simple computer program, written in BASIC, is given for the processing of photoelectric data on personal computers. Finally, we suggest several ways in which the observations can be preserved and archived so that they will be useful to the astronomical community now and in the future.

II. SELECTION OF AN OBSERVING PROGRAM

Begin by selecting a scientifically worthwhile program that is of interest to you and is compatible with your equipment, observing site, time, and ability. It is useful to consult with the A.A.V.S.O., the I.A.P.P.P., or professional astronomers for advice in selecting an observing program and carrying it out. Several interesting observing programs are discussed in this book and in the *A.A.V.S.O. Photoelectric Newsletter* and the *I.A.P.P.P. Communications* over the last few years. Useful photometry can be done with small telescopes or at sites having bright skies if bright stars are selected for study. Participation in observing programs with others and in international observing campaigns is very satisfying and often exciting. Of current interest and need are the photometry of Be stars, short-period pulsating variables, eclipsing binaries, long period yellow and red variables, and RS Canum Venaticorum variables. Photometry of novae and flare stars, as well as asteriods, and occultations of stars by asteriods, comets, and planets are also interesting.

III. READING THE LITERATURE

Once you have decided on an observing program, it is valuable to learn as much as possible about the stars that you are observing as a group (*e.g.*, RS CVn variables) and on an individual basis (*e.g.*, λ And). It is useful to collect material on the objects that you are observing and keep it in a folder labelled with the star's name. In this way a "file" or dossier can be built up on each star. Learning more about the stars that you are observing will make your observing project more enjoyable and more intellectually rewarding. It certainly helps to know why a certain star is being observed, especially in the wee hours of a winter morning when you are cold and tired.

IV. INSTRUMENTATION

It is important to maintain the telescope and the photometric equipment in good operating condition. The optical surfaces of the telescope and photometer should be kept clean and the detector and electronic equipment should be checked frequently. It is also important that the telescope is properly aligned and that the drive can keep the star in the centre of the diaphragm for at least a few minutes. Make calibrations and checks of the equipment at least twice a year and keep a record in a log book. Where applicable, careful calibrations of amplifier gains or the determinations of "dead time" constants also should be made. Keep a record of the filters, detector, amplifiers, and power supplies that are used and any modifications or changes that are made, even if they are small changes. If you can afford the expense, keep duplicates of electronic and optical parts. This helps in trouble-shooting and leads to a minimum of down-time. In addition, one common, but easily correctable, problem is faulty or loose electrical connections which can produce erratic or intermittent fluctuations in the output signal. Electrical connections to and from the detector are especially important and should be as tight as possible.

V. OBSERVING TECHNIQUE

No amount of data processing either by hand or by sophisticated computer program can salvage bad photometric data. The poor quality of photometric observations can arise from a variety of causes, but the most common causes are the presence of clouds, faulty equipment, or bad observing technique. Strive for the greatest accuracy you can attain. Don't observe on cloudy nights. To quote John Percy in this matter of faith and photometry, "A dubious observation is worse than no observation at all". High, thin cirrus clouds are difficult to see on a dark night. This cloud type is associated with an advancing warm front and generally does not move away rapidly. It is important to make a note of the sky conditions during evening and morning twilight periods when these high clouds are usually easily visible. If they are seen at the start and end of a night, it is probable that these clouds were around throughout the entire night, and that the measurements made on that night are suspect. Another common source of observational error is not centering the star in the diaphragm. If the star drifts out of the centre of the diaphragm, errors of $\sim 1\% - 2\%$ may result due to sensitivity changes across the photocathode surface.

Observe differentially with respect to a comparison star which is close to the variable star in angular separation. The rule of thumb is that the variable and comparison stars should be *less* than a degree apart. This is to reduce the uncertainties in the atmospheric extinction corrections. In addition, the apparent magnitude and colour index of the comparison star should be as close as possible to those of the variable star. A difference in brightness between the variable and comparison of ≤ 1.5 mag. is usually sufficient to keep possible errors arising from non-linearities in the detector/amplifier system, uncertainties in amplifier gain calibrations, coincidence corrections, or chart recorder read-off errors to within 0.005 mag. If very bright stars ($V \leq +3.0$ mag) are observed, it may be advisable to stop down the aperture of the telescope or to use a neutral density filter to reduce the incoming light so that it doesn't saturate the detector. Also selecting a comparison star with a $B - V$-index close to that of the variable ($\Delta(B - V) \leq 0.3$ is usually adequate) minimizes second-order extinction corrections and reduces uncertainties in transforming the observations to a standard photometric system.

What is most important, however, is that the comparison star be *constant* in brightness. The constancy in brightness of the comparison star should be verified by observing a check star. It is also worthwhile to consult the *General Catalogue of Variable Stars* (Kholopov 1985) and *The New Catalogue of Suspected Variable Stars* (Kholopov 1982) to ascertain that the comparison and check stars are not listed as variables or suspected variables. If the comparison star is brighter than 6th magnitude, there may be a note on its variability in the *Bright Star Catalogue* (Hoffleit 1982).

Our own experience indicates that almost all luminous supergiants (those with I, Ia, Iab luminosity types) and M-stars are variable. Also be wary of O and early B stars since a significant percentage of these stars also show small brightness variations. These stars should not be used as comparison or check stars. In addition, unless there is no other suitable choice, one should avoid using binary stars as comparison or check stars since you are doubling the risk that one of the stars will be a variable.

Follow the usual pattern of *sky-comparison-variable-(check)-comparison-sky* and try not to spend more than $\sim$ 10 minutes between successive comparison star measures. The comparison star observations are just as important as the variable star observations in differential photometry. Typically about 30-40 seconds of time should be spent on each comparison and sky measure. For faint stars, longer observing times may be necessary. Also for faint stars and crowded star fields, be sure to check that *no* unwanted stars are visible in the field of the diaphragm. It is good practice to take the sky at the same fixed location for a particular variable star. Typical diaphragm apertures range from 30-60 arcseconds. Of course, the diaphragm size should not be changed when intercomparing the variable star and its comparison.

When sky brightness is rapidly changing due to the rising or setting of a bright moon or when working in the twilight periods, it may be necessary to observe the sky more frequently. Also try to avoid observing at large values of air mass (*arcsec* z). Values of *arcsec* $z \geq 2.00$ (*i.e.*, zenith angles of $\geq 60°$) should be avoided unless there are compelling reasons to observe a particular star through a large air mass. At large air masses, the extinction corrections become significant and, of course, there is a higher likelihood that the star's light path will intercept clouds or haze. More information on making accurate photoelectric observations is given by Hall in this book and by Hall & Genet (1982).

VI. STANDARD PHOTOMETRIC SYSTEMS

Today there are over a dozen photometric filter systems in use for astronomical photoelectric photometry (*e.g.*, see Strömgren 1963; Hendon & Kaitchuck (1982) for discussions). However, the two most widely used photometric systems are the Johnson & Morgan (1953) wideband ($\Delta\lambda \simeq 700 - 900$ Å) *UBV* system and the Strömgren (1963) intermediate band ($\Delta\lambda \sim 200 - 300$ Å) *uvby* system. Although in many respects the Strömgren *uvby* system is superior to the *UBV* system (see Strömgren 1963), the *UBV* system is preferred by most people working with small telescopes chiefly because the broad bandpasses of the filters allow more photons to reach the detector. In the Strömgren *uvby* system, much of the incoming light is filtered out before reaching the detector. As discussed in Section VII, however, the very broad bandwidths of the *UBV* systems have a disadvantage in not being sensitive to important stellar features. Also, the spectral response curves of the bandpasses are dependent on the spectral sensitivity of the detector, as well as on the spectral energy distribution of the source. Despite these disadvantages, the widespread use of the *UBV* system and the large number of stars that have been observed using it, enhance its appeal to many photometrists to the point where it has become the *de facto* standard "international" photometric system. In addition, with the increased spectral sensitivity of photoelectric detectors to red light, the *UBV* system has been extended to include additional bandpasses such as the red R ($\lambda_{eff} \sim 7000$ Å) and (near) infrared I ($\lambda_{eff} \simeq 9000$ Å) bandpasses to obtain the *UBVRI* system.

Special-purpose photometric systems also have been developed for isolating and measuring certain important spectral features in stars. Probably the most widely used for variable star work are the Beta system of Crawford (1958) and the Alpha system (*e.g.*, Baluinas, Ciccone & Guinan 1975) which consist of narrow band ($\Delta\lambda \simeq 30 - 35$ Å) and intermediate band ($\Delta\lambda \simeq 250 - 300$ Å) interference filter pairs, centered near the rest wavelength of the hydrogen Balmer Hβ ($\lambda 4861$) and Hα ($\lambda 6563$) features, respectively. These photometric systems are particularly useful for studying stars with strong hydrogen line emissions such as Be stars (*e.g.*, Guinan & Hayes 1984), interacting close binaries (see Guinan *et al.* 1975), and RS CVn, BY Dra, and FK Com type variables which show Hα emission. Figure 1, from Dorren, Guinan, & McCook (1984), shows the relatively large enhancements (flares?) in the

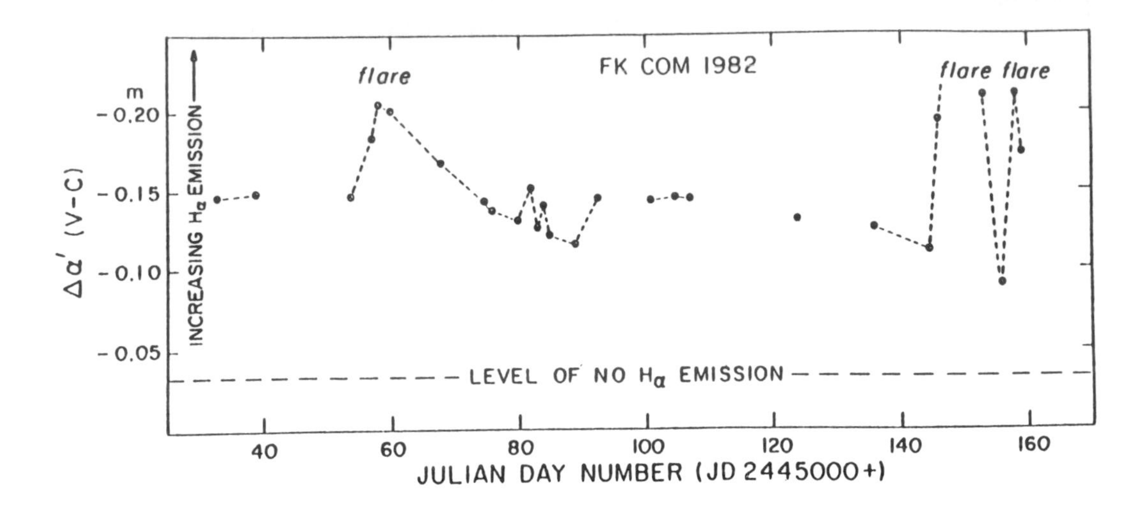

Figure 1. The differential alpha index ($\Delta\alpha'$), in the sense variable star minus comparison star, is plotted against the Julian Date for the rapidly rotating G giant FK Comae. More negative values of α indicate stronger net Hα line emission. The Hα flare events and the level of no net Hα emission are indicated in the figure. (From Dorren, Guinan, & McCook 1984.)

Hα line emission strength, as measured by Hα photometry, for the rapidly rotating G giant FK Comae.

In addition, a useful photometric system has been developed by Wing for studying cool stars (see article by him elsewhere in this volume). It is especially suitable for the study of the physical changes that Mira stars undergo during their pulsational cycle.

VII. THE *UBV* SYSTEM: STANDARDIZATION PROBLEMS

Although there are a number of standard photometric systems in use today for photoelectric photometry, the *UBV* system is the most widely used. The properties of the *UBV* system have been thoroughly described in a number of papers (*e.g.*, Johnson & Morgan 1953; Johnson 1963; Becker 1963) and will be discussed here with regard to transforming observations made with filters that match the *UBV* system to the standard system. A summary of the spectral characteristics of the *UBV* system is given in Table 1 along with comments relevant to the dependency of the effective wavelength and bandwidth to the spectral response of the detector and the source. The RCA 1P21 (S4) photomultiplier, which was used by

Table 1. Properties of the *UBV* system.

Designation	Filters	λ_{eff} (Å)*	$\Delta\lambda$ (Å)*	Comments
U	Corning No. 9863	3660	~ 700	Short wavelength cutoff determined by atmospheric attenuation in the UV and by telescope optics; λ_{eff} strongly dependent on the colour and strength of the Balmer discontinuity at λ3646; also "red leak" problems.
B	Corning No. 5030 +2 mm Schott GG-13	4400	~ 970	λ_{eff} is a function of spectral type of source; long and short wavelength cutoffs defined chiefly by filters; possible "red leak" problems with red sensitive detectors.
V	Corning No. 3384	5530	~ 850	Long wavelength cutoff determined by red sensitivity of the detector; for M stars, TiO bands within the bandpass strongly affect the flux.

* The effective wavelengths and half-widths $\Delta\lambda$ are from Schmidt (1956). An RCA type 1P21 PMT (S4 surface) and a reflector telescope with aluminized mirrors are assumed.

Johnson & Morgan in establishing the UBV system, has essentially no sensitivity to light with wavelengths longward of ~ 6300 – 6500 Å (see Figure 2). However, the spectral sensitivity of many modern photoelectric detectors extends out to wavelengths near 10,000 Å. Filter combinations different from those used originally by Johnson & Morgan must therefore be used with red sensitive detectors because the original filters are not blocked to red light. A set of glass filters which, when used with red sensitive detectors, satisfactorily matches the UBV system is given by Bessell (1976). These or similar filter sets should be used with red sensitive detectors so that the observations can be transformed to the standard system. Filters that closely reproduce the spectral responses of the R and I bandpasses are also given by Bessell.

The chief advantage of observing variable stars with a small telescope in the UBV system is that the bandpasses are very broad, which allow many photons to pass through the filter to the detector. In addition, photometry in the UBV system, once it is standardized, allows the observations to be compared and/or combined with other observations made on the system. Although the broad bandwidths of the UBV system have the advantage of making larger numbers of fainter stars accessible to photometry, the broad bandwidths have the disadvantage of being too wide to isolate certain astrophysically important spectral features — such as emission lines. In addition, the effective wavelength of each regional band is not fixed for a particular telescope-photodetector-filter combination, but is dependent also on the spectral energy distribution of the source (the star) and to a lesser degree on the effects of interstellar and atmospheric extinction. This is an unfortunate situation and means that the net spectral response of the filter-system changes with the colour of the source, in the sense that the response functions of the bandpasses shift toward shorter wavelengths for blue (=hot) stars and toward longer wavelengths for red (=cool) stars (see Johnson 1955). For example, Figure 2 illustrates schematically the factors that go into the final net spectral response curve and the effective wavelength of the V bandpass. The B and U bandpasses show an even stronger dependency on the spectral energy distribution of the source.

The behaviour of the U bandpass is more complex because its spectral response is greatly influenced by the spectral feature that it was originally designed to measure, *i.e.*, the hydrogen Balmer discontinuity at ~ 3646 Å. This feature is extremely strong for late B- and early A-type stars in which flux shortward of 3660 Å is greatly reduced by continuous absorption by hydrogen.

The spectral response curves for the UBV filter system have been calculated by Spear (1971) using stellar spectrophotometric intensity distributions of stars with a wide range of spectral types. Figure 3 shows the non-linear behaviour of the effective wavelength of the U bandpass as a function of spectral type and $(B-V)$ as found by Spear. This study shows that the effect of the colour of the star on the

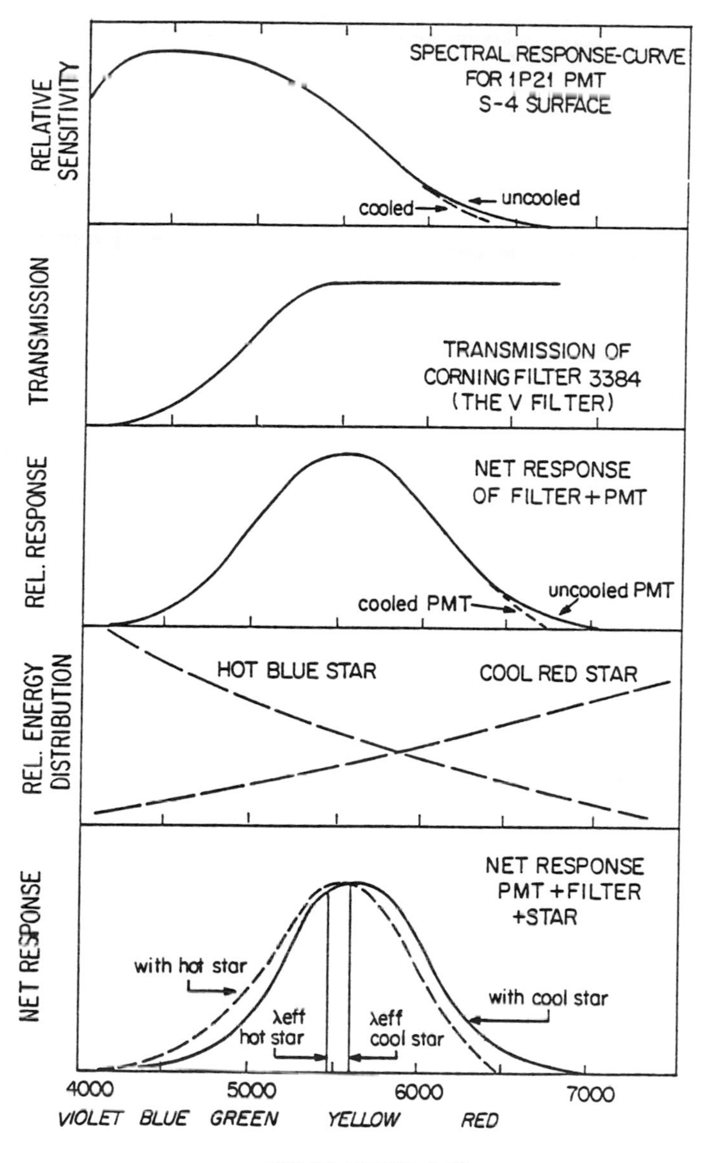

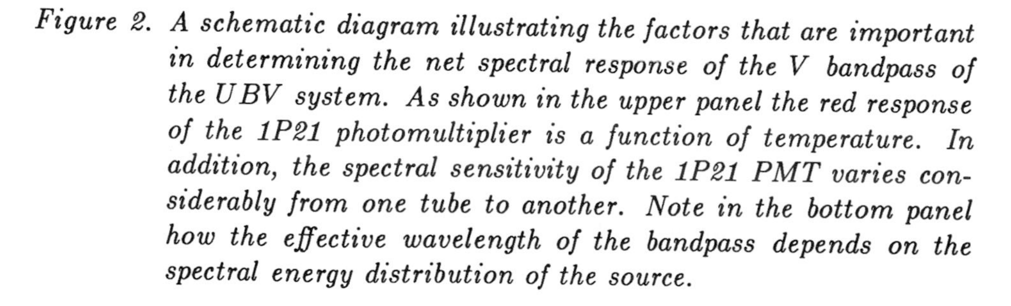
Figure 2. A schematic diagram illustrating the factors that are important in determining the net spectral response of the V bandpass of the UBV system. As shown in the upper panel the red response of the 1P21 photomultiplier is a function of temperature. In addition, the spectral sensitivity of the 1P21 PMT varies considerably from one tube to another. Note in the bottom panel how the effective wavelength of the bandpass depends on the spectral energy distribution of the source.

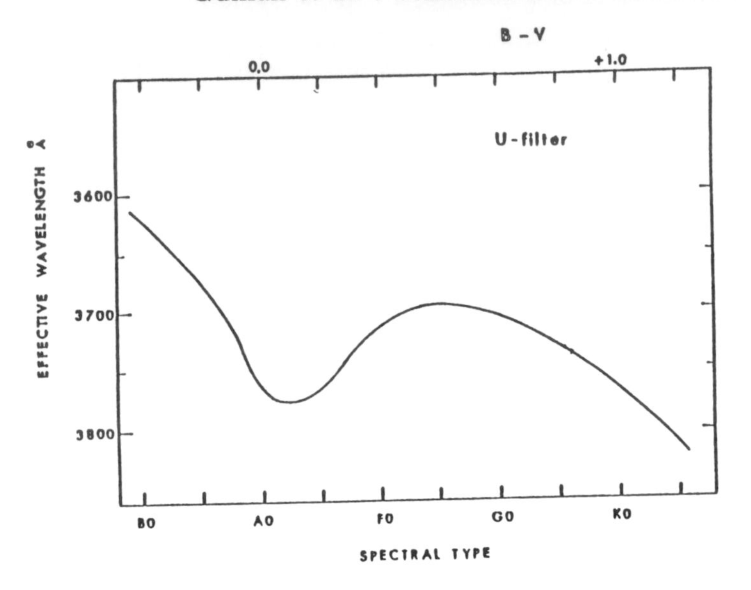

Figure 3. Variation of the effective wavelength of the U bandpass with spectral type and $B-V$.

effective wavelength is relatively small for the V bandpass but is significant for the B and U bandpasses.

For the reasons discussed above, the UBV system is a difficult photometric system to reproduce and to standardize with high precision. In addition, the variation of the effective wavelengths of the bandpasses with the star's colour and the wavelength dependence of the extinction of the Earth's atmosphere introduce a colour-dependent term in the measurement of atmospheric extinction of the form: $k_\lambda = k'_\lambda + k''_\lambda(B-V)$. Here k_λ is the measured extinction coefficient for a star with a particular $(B-V)$-index and k'_λ and k''_λ are the so-called principal and colour-dependent extinction terms respectively. The presence of the k''_λ term is a result of the dependency of the UBV bandpasses on the colour of the star. Generally, k''_V is very small, -0.02 mag $\leq k''_V \leq 0.00$ mag, while the colour term for the B bandpass is typically -0.05 mag $\leq k''_B \leq -0.02$ mag. The colour-dependent extinction term for the U bandpass is more complicated because of the presence of the Balmer discontinuity within its bandpass. However, the original UBV system was set up on the erroneous (but practical) assumption that $k''_U = k''_B$, so that in transforming to the standard system, this error is retained and $k''_U = k''_B$. Following the method of Hardie (1962), the colour-dependent terms can be determined empirically by measuring the extinction coefficients of two stars of different colours that form a close pair. The measured extinction coefficients should be slightly different (smaller for the redder star of the pair). The extinction determination is done in the usual manner by observing the star pair over a large range of air mass. Once the k'' values are determined, they remain essentially fixed with time for a particular telescope-filter-detector system. A list of several bright red-blue star pairs is given in Table 2.

The value of the k'' term can be found for the red/blue close star pair by using the formula

$$k'' = \frac{k_1 - k_2}{[(B-V)_1 - (B-V)_2]} \tag{1}$$

in which k_1 and k_2 are the measured extinction coefficients for star 1 and 2 respectively, while $(B-V)_1$ and $(B-V)_2$ are the corresponding colours.

The computer program given at the end of the paper can be used to determine k'' by encoding the two stars as variable and comparison and using the least squares extinction determination to find k for that star and then reverse the coding of the variable and comparison and calculating the value of k for the other star. Typical values for atmospheric extinction coefficients for the UBV bandpasses for sites with elevations less than about 1000 feet are: $k_U \sim 0.65$ mag, $k_B \sim 0.38$ mag, and $k_V \sim 0.22$ mag.

Taking V of the UBV system as an example, the instrumental (unstandardized) differential magnitude $\Delta V'_\circ$, corrected for atmospheric extinction is given by:

$$\Delta V'_\circ = \Delta V' - k'_V\ \Delta X - k''_v\ \overline{X}\Delta(B-V) \tag{2}$$

where $\Delta V'$ is the measured differential magnitude of the two stars, ΔX is the difference in the air mass between the two stars (or $\Delta X = arcsec\ z_1 - arcsec\ z_2$), $\overline{X}$ is the mean air mass value of the two stars (or $\overline{X} = (X_1 + X_2)/2$), and $\Delta(B-V) = (B-V)_1 - (B-V)_2$.

The equation for transforming the extinction-corrected, instrumental differential magnitude $\Delta V'_\circ$ to ΔV of the UBV system is

$$\Delta V = \Delta V'_\circ + \epsilon_V \Delta(B-V) \tag{3}$$

in which ϵ_V is the transformation coefficient. As discussed by Hardie (1962) and by Hall & Genet (1982), the value of the transformation coefficient can be determined easily and accurately by using the blue/red close star pairs again. In this case, it is presumed that the colour-dependent (k''_V) extinction coefficient has been determined and that the value of $k'_V \Delta X$ is very small due to the small angular separation of the two stars of each pair. To make ΔX as small as possible for a given star pair, the observations should be made close to the meridian. Thus from equation (3), the value of the transformation coefficient is

$$\epsilon_V = \frac{\Delta V - \Delta V'_\circ}{\Delta(B-V)} \tag{4}$$

Table 2. Star sets for UBV standardization and calibration.

HR	NAME	RA(2000)			DECL(2000)			Sp.	V	$B-V$
		h	m	s	°	'	"			
458	υ And	01	36	48	+41	24	20	F8 V	+4.09	+0.54
477	τ And	01	40	35	+40	34	37	B8 III	4.94	−0.09
483		01	41	47	+42	36	49	G1.5 V	4.95	+0.61
1051		03	30	37	+48	06	13	B8 V	5.82	−0.04
1052	σ Per	03	30	34	+47	59	43	K3 III	4.36	+1.35
1226		03	58	29	+38	50	25	gK1	6.30	+1.08
1229		03	59	40	+38	49	14	A1Vp	6.38	+0.10
5854	α Ser	15	44	16	+06	25	32	K2 III	2.65	+1.17
5859		15	45	23	+05	26	49	A0 V	5.58	+0.04
6418	π Her	17	15	03	+36	48	33	K3 II	3.16	+1.44
6436	69 Her	17	17	40	+37	17	29	A2 V	4.65	+0.05
8451		22	10	21	−03	53	39	A1 V	6.27	−0.01
8453		22	10	34	−04	16	01	K0 III	6.01	+0.98

or more generally,

$$\epsilon_V = \frac{\Delta V - \Delta V' + k'_V \Delta X + k''_V \overline{X} \Delta(B-V)}{\Delta(B-V)} \qquad (5)$$

where the values of ΔV and $\Delta(B-V)$ are known, for example from Table 2, and where the other quantities are measured or directly computed. It is wise to use at least three or four of the blue-red star pairs when determining the transformation coefficients. The average of the determinations should then be used.

As suggested earlier, if the colours of the variable and comparison stars are close (*i.e.*, withing ~ 0.30 in $\Delta(B-V)$), then uncertainties in the determination of the transformation coefficient will result in negligibly small errors in determining ΔV or ΔB. For example, if the true value of $\epsilon_V = -0.02$ and $\epsilon_V = -0.04$ is measured, then for $\Delta(B-V) = 0.30$ and $\Delta V'_\circ = 1.00$, and $\Delta V_{true} = 0.994$ mag while the value of $\Delta V = 0.988$ mag would be calculated using the erroneous value of ϵ_V. The computer program will standardize the differential magnitudes in the manner described above.

VIII. REDUCTION OF PHOTOELECTRIC OBSERVATIONS

The photoelectric measurements of the sky, comparison, check, and variable star made at the telescope, along with the recording of the times at which the observations were made, constitute the "raw" photometric data base. To make these data useful for astronomical purposes, it is necessary to compute the differential magnitudes of the variable star relative to the comparison star which are corrected for atmospheric extinction and preferably transformed to a standard photometric system such as the *UBV* system. In addition, the observed local time must be converted to heliocentric Julian Date (HJD) so that the varying light travel time from the star due to the Earth's orbital motion around the sun can be eliminated. Heliocentric time is the time when the light signal from the star reaches the sun, and the maximum heliocentric correction is $\pm 8^m 19^s$.

For a star that varies in brightness periodically such as an eclipsing binary or a periodic pulsating star, it is useful to compute its phase. This can be done quite easily from the heliocentric Julian Date, using the appropriate light elements for the variable star. The details about the reduction of photometric observations are given by Wood (1963), Hardie (1962), and more recently by Hall & Genet (1983).

Photoelectric photometry of variable stars has been carried out at Villanova University since 1967. Since that time, the photometric observations have been reduced by computer. In the last few years, the increased speed and storage capacity of personal computers have made it not only possible but practical to process and reduce photometric observations. Moreover, the use of a microprocessor-controlled data acquisition and recording system permits the data to be read directly onto a magnetic tape or floppy disk and processed by computer. Even if the observations are read from a chart recorder, a digital meter, or printed form, the data can be manually encoded on a floppy disk and also reduced by computer. Editing and correcting the data can be done conveniently on a video terminal or from printed output. A computer reduction program for differential photoelectric photometry similar to the large program in use at Villanova is given here. Some modifications to the reduction program will certainly be necessary to accommodate the various data acquisition and recording systems now in use. Because the program is written in BASIC to run specifically on an IBM PC, running it on other computer systems may require some changes to be made. There are numerous comment or remark (REM) statements included in the computer program which will make it easier to implement the necessary changes and additions. The Julian Date and sidereal time are calculated internally by the computer program from the civil data and local standard time, using formulae given in the *Almanac for Computers*, published by the Nautical Almanac Office of the U.S. Naval Observatory. The conversion of Julian Date to heliocentric Julian Date (HJD) is also accomplished from solar coordinates generated from formulae given in the *Almanac for Computers*. The formulae used in the present program are for 1985 and 1986, and these should be updated for future years. When using the computer program, don't forget to enter your values for longitude, latitude, and time zone in the main program.

A sample data set is provided along with the corresponding reduced output. It is advisable to run these test data to be certain that the program statements and commands have been properly entered on the computer. It is also possible to get a copy of the program on a floppy disk (IBM diskette; standard format) by writing to the authors.

IX. ANALYSIS AND INTERPRETATION OF PHOTOMETRIC DATA

The ultimate goal of astronomical photoelectric photometry should be to learn more about the physical properties of the objects being observed. In the case of variable stars, the light variations carry important information on the physical processes responsible for them. As Zdenek Kopal discusses in his book *Language of the Stars*, the interpretation and analysis of the light changes of variable stars is a type of astronomical cryptography, in which the job of the analyst is to "decode" the photometric data to yield the information which it contains. Although Kopal was referring to the determination of the orbital and physical properties of eclipsing binary star systems obtained from an analysis of their light curves, it also applies to pulsating variable stars, RS CVn variables as well as other types. To understand the physical cause or causes of the light variations, it is necessary to develop models based on solid physical theory so that we can decipher what the observations have to say, and extract as much information as possible about the nature of the objects under study. Examples of the kinds of research and analysis techniques that are used in the study of several commonly observed types of variable stars are given in Table 3.

It is, however, beyond the scope of this paper to discuss in detail methods of data analysis and modelling that are currently utilized in the study of variable stars. More information on the theory and analysis of photometry of variable stars is given in the following books: *Properties of Double Stars* (Binnendijk 1960); *Photoelectric Astronomy for Amateurs* (Wood 1963); *Variable Stars* (Strohmeier 1972); *Interacting Binary Stars* (Sahade & Wood 1978) and *Variable Stars* (Hoffmeister, Richter, & Wenzel 1985). The last reference provides (on pp. 295-298) a comprehensive list of the important general summaries, compilations, review articles, conference proceedings and monographs that deal with almost every known class of variable star. In addition, the book *Software for Photometric Astronomy* by Ghedini gives a number of computer programs that are useful for the analysis of variable star observations.

X. PUBLICATION OF THE RESULTS

With the patience, time, and effort required to obtain good quality photoelectric data, it behooves the observer to make the observations known to the scientific community now and for the future. Having photoelectric measurements stored away in a desk drawer and forgotten is like having no data at all! The observations should be listed in tabular form, preferably with the time of the observations given in heliocentric Julian Date (HJD) and the star's brightness expressed in a standardized photometric system such as the *UBV* system as ΔV, ΔB, etc., if appropriate. If the data were obtained with filters matching those used in a standard photometric system such as the *UBV* system, but *not* standardized, one should clearly indicate this by using $\Delta V'$ or $\Delta Y'$ designations and *not* ΔV! If mean points are given, it is useful to know how many observations constitute a mean and also indicate the precision of the observation by giving its standard deviation. The following offers a few suggestions for having the data preserved or published.

a) Publish a paper on the star in an astronomical publication such as the *Astronomical Journal, Publications of the Astronomical Society of the Pacific, Information Bulletin on Variable Stars, Journal of the A.A.V.S.O.*, or *I.A.P.P.P. Communications.* Include the data with the paper in tabular form. Some journals do not accept large tabulations of data, however. It

Table 3. Photometric analyses of some commonly observed variable stars.

Variable Star Type	Physical Mechanism(s) of Variability	Analysis	Scientific Objectives
Eclipsing Binaries			
a) times of minimum light	Mutual eclipses of the binary components	Mathematical or graphical determination of the time of mid-eclipse	Variation of the orbital period with time: period studies yield information on mass loss and exchange
b) total light curves	Light variations produced by eclipses and by tidal and reradiation effects of the components	Analysis of the full light curve using Fourier methods or light curve synthesis computer programs	Determination of the orbital elements and the physical properties of the stars
RS CVn/BY Dra and related variables	Light modulation produced by dark starspots	Analysis of multi-wavelength light curves with spot model programs or with Fourier transform techniques	Determination of the starspot properties: spot temperature area, number and distribution; also variations of these with time; differential rotation and spot cycles
Be stars	Ejection of shells and disks of gas; (non-radial pulsations also occur)	Period searches; short term and long term behaviour of brightness, colour, and spectra	Mechanism(s) operating; stellar instabilities; determination of the mass loss
Slowly-varying pulsating stars (Miras)	Pulsation; shock mechanisms in atmosphere	Analysis of light and colour changes with time	Determination of the pulsation mechanisms operating; stellar structure
Short period variables (β Cep; δ Sct variables)	Pulsational instabilities; non-radial pulsation modes	Fourier analyses; discrete Fourier transforms; other period analyses	Determination of multiple periods; non-radial pulsations; internal structure/evolution

is helpful to read papers written on a similar type of star and model your paper on that, or collaborate or get advice from someone who has published many papers.

b) The International Astronomical Union (I.A.U.) maintains an astronomical data archival service to which data can be deposited (see Breger 1985). The individual differential magnitudes can be sent to the I.A.U. Commission 27 Archive for Unpublished Observations of Variable Stars. Three copies of the observations as well as a brief descriptive cover sheet should be submitted to the current coordinator of the Archive. The data files can be hand written in black ink or typed, and the printed part should be no larger than 20 by 27 cm per sheet. An identification file number will be assigned to the data and the title listings are published yearly. The address of the current Coordinator is:

Dr. Michel Breger
Coordinator of IAU Archives
Universitäts-Sternwarte
Türkenschanzstr. 17
A-1180 Wien
Austria

The PEP section of the A.A.V.S.O. also accepts photoelectric data on variable stars. The A.A.V.S.O. also serves as a liaison between those having photoelectric measurements of a particular class of variable with others who are working on those stars or who need photometry as part of their study. Contact the A.A.V.S.O. for more information.

REFERENCES

Baluinas, S.L., Ciccone, M.A., & Guinan, E.F. (1975). *Pub. A. S. P.*, **87**, 969.

Becker, W. (1963). In *Astronomical Techniques*, ed. K. Aa. Strand, p. 241. Chicago: University of Chicago Press.

Bessel, M.S. (1976). *Pub. A. S. P.*, **88**, 557.

Binnendijk, L. (1960). *Properties of Double Stars.* Philadelphia: University of Pennsylvania Press.

Breger, M. (1985). *Pub. A. S. P.*, **97**, 85.

Crawford, D.L. (1958). *Ap. J.*, **128**, 185.

Dorren, J.D., Guinan, E.F., & McCook, G.P. (1984). *Pub. A. S. P.*, **96**, 250.

Ghedini, S. (1982). *Software for Photometric Astronomy.* Richmond: Willmann-Bell.

Guinan, E.F. & Hayes, D.P. (1984). *Ap. J.*, **279**, 721.

Guinan, E.F., McCook, G.P., Bachmann, P.J., & Bistline, W.G. (1976). *A. J.*, **81**, 57.

Hall, D.S. & Genet, R.M. (1982). *Photoelectric Photometry of Variable Stars.* Fairborn: I.A.P.P.P.

Hardie, R.H. (1962). In *Astronomical Techniques*, ed. W.A. Hiltner, p. 178. Chicago: University of Chicago Press.

Henden, A.A. & Kaitchuck, R.H. (1982). *Astronomical Photometry.* New York: Van Nostrand Reinhold.

Hoffleit, M. (1982). *Bright Star Catalogue.* New Haven: Yale University.

Hoffmeister, C., Richter, G, & Wenzel, W. (1985). *Variable Stars.* Berlin: Springer-Verlag.

Johnson, H.L. (1955). *Ann. d'Ap.*, **18**, 292.

Johnson, H.L. (1963). In *Basic Astronomical Data*, p. 204. Chicago: University of Chicago Press.

Johnson, H.L. & Morgan, W.W. (1953). *Ap. J.*, **117**, 313.

Kopal, Z. (1979). *Language of the Stars.* Dordrecht: Reidel Publ. Co.

Kholopov, P.N. (ed.) (1982). *New Catalogue of Suspected Variable Stars.* Moscow: Nauka Publ. Office, U.S.S.R.

Kholopov, P.N. (ed.) (1985). *General Catalogue of Variable Stars*, 4th Edition. Moscow: Nauka Publ. Office, U.S.S.R.

Sahade, J. & Wood, F.B. (1978). *Interacting Binary Stars.* Oxford: Pergamon Press.

Schmidt, K.H. (1956). *Mitt. Obs. Jena*, No. 21.

Spear, G. (1971). *Ph.D. Thesis*, University of Pennsylvania, Philadelphia, Pa.

Strohmeier, W. (1972). *Variable Stars.* Oxford: Pergamon Press.

Strömgren, B. (1963). In *Basic Astronomical Data*, ed. K.Aa. Strand, p. 123. Chicago: University of Chicago Press.

Wolpert, R.C. & Genet, R.M. (1983). *Advances in Astronomical Photometry*, Vol. 1. Fairborn, Ohio: Fairborn Observatory Press.

Wood, F.B. (1963). *Photoelectric Photometry for Amateurs.* New York: Macmillan.

```
10 REM *   DATA REDUCTION AND STANDARDIZATION PROGRAM FOR DIFFERENTIAL    *
11 REM *                                PHOTOMETRY                        *
20 REM *   INSTRUCTOR:DR MC COOK            WRITTEN BY JOSEPH MC MULLIN   *
21 REM THIS PROGRAM SETUP FOR B,V OBSERVATIONS   MAY BE MODIFIED FOR
22 REM ADDITIONAL FILTERS ETC.
23 REM IF USING PULSE COUNTING PHOTOMETERS, DON'T FORGET TO INCLUDE 'DEAD
24 REM TIME' CORRECTION
25 REM INCLUDES PRIMARY AND COLOR DEPENDENT ATM. EXTINCTION TERMS
30 DIM AR(8),X(270),Y(270),CODE(270),DEFL(4,270),TIME(4,270),D$(270)
40 REM COORDINATES OF THE OBSERVATORY THESE MAY BE ADJUSTED ACCORDINGLY
50 REM LATTITUDE OF THE OBSERVATORY(CONVERTED FROM DEGREES TO RADIANS)
60 LAT=(29+37/60+59.7/3600)*3.14159/180:REM LAT=(DD+MM/60+SS.S/3600)
70 REM LONGITUDE OF THE OBSERVATORY
80 LONG=(-52+31/60)/15:REM LONG=(DD+MM/60):U=1:V=1
90 DEF FNARCSIN(X)=ATN(X/SQR(1-X*X))
91 REM THIS IS THE FORMAT THAT INFORMATION IN THE INPUT FILE SHOULD BE
92 REM ARRANGED
95 REM 081077SH.DAT  FILENAME.FILETYPE(MONTH(MM),DAY(DD),YEAR(YY)INITIALS).DAT
96 REM XPER          VARIABLE STAR NAME
97 REM 03,53,56.8    RA OF VARIABLE STAR
98 REM 30,58.9       DECLINATION OF VARIABLE STAR
99 REM HR1191        COMPARISON STAR NAME
100 REM 03,50,29.5   RA OF COMPARISON STAR
101 REM 34,17        DECLINAITON OF COMPARISON STAR
102 REM SHIRAZ       OBSERVER'S NAME
103 REM -0.01        EPSILON V
104 REM -0.03        EPSILON B
105 REM  0.30        B-V OF VARIABLE
106 REM  0.10        B-V OF COMPARISON
107 REM  0.22        K PRIME(V)
108 REM  0.40        K PRIME(B)
109 REM -0.01        K SECONDARY(V)
110 REM -0.03        K SECONDARY(B)
111 REM 2443055.0#   EPHEM:E.G.TIME OF MAX (# INDICATES DOUBLE PRECISION)
112 REM 620.0#       PERIOD(OF THE VARIABLE STAR IN DAYS)
113 REM SYSTEM       SYSTEM(THIS WORD SIGNALS THE BEGINNING OF DATA STRINGS)
114 REM 140000279:034700   DATA STRING
115 REM DATA STRINGS ARE BROKEN UP INTO VARIOUS SECTIONS:
116 REM COLUMN 1 INDICATES THE FILTER USED,COLUMN 2 INDICATES THE SKY CODE
117 REM COLUMNS 3 THROUGH 5 ARE PADDED ZEROES(SO THE SYSTEM IS COMPATIBLE
118 REM WITH VILLANOVA COMPUTER GENERATED DATA AS WELL AS CHART DATA)
119 REM COLUMNS 6 THROUGH 9 ARE THE ACTUAL DEFLECTION READINGS
120 REM COLUMNS 11 THROUGH 16 INDICATE TIME (LOCAL) OF READING(HHMMSS)
121 REM STOP         STOP(THIS WORD SIGNALS THE END OF DATA STRINGS AND
122 REM                     THE END OF INPUT)
123 REM                              NOTE
124 REM    IF THE B-V VALUES OF THE COMPARISON AND VARIABLE ARE UNKNOWN,
125 REM   PROVISIONAL VALUES SHOULD BE ASSUMED FROM THE SPECTRAL TYPES OF
126 REM    THE STARS.  AFTER  THE FIRST RUN, THE ACTUAL VALUES OF B-V
127 REM   FOR EACH STAR MAY BE OBTAINED FROM THE COMPUTER PROGRAM AND
128 REM              INPUT INTO A REVISED INPUT FILE
129 REM NOTE : PROGRAM DOES NOT ACCOUNT FOR USE OF A CHECK STAR
130 REM        PROVISIONS FOR A CHECK STAR MUST THUS BE ADDED
135 PRINT "Load disk into drive  ***** enter PC drive and file name (ie b:fn.ft)"
140 INPUT FILN1$
150
160 OPEN FILN1$ FOR INPUT AS #2
170 M#=VAL(MID$(FILN1$,3,2))
180 DAY#=VAL(MID$(FILN1$,5,2))
190 YEAR#=VAL(MID$(FILN1$,7,2)):YEAR#=1900#+YEAR#
```

```
200 IF YEAR#=1986 THEN GOSUB 4370:GOTO 310
202 REM SIDEREAL TIME CALCULATION CONSTANTS FROM ALMANAC FOR OBSERVERS
204 REM THESE CONSTANTS ARE ACCURATE FOR THE YEAR 1985 AFTER WHICH THEY
206 REM                      MUST BE ADJUSTED
210 AR(1)=37.46159684#
220 AR(2)=12.02492339#
230 AR(3)=5.97E-06
240 AR(4)=1.95E-06
250 AR(5)=5.48E-06
260 AR(6)=1.694E-05
270 AR(7)=-4.96E-06
280 AR(8)=-8.16E-06
290 A=183
300 W=1
310 REM
320 INPUT #2,DUMMY$
330 PRINT:INPUT #2,V$
340 PRINT:INPUT #2,VRA#,MRA#,SRA#
350 RAV#=VRA#+MRA#/60#+SRA#/3600#
360 PRINT:INPUT #2,VDEC#,MDEC#
370 IF VDEC#<>ABS(VDEC#) THEN DECV#=VDEC#+MDEC#/60#:DECV#=-DECV#:GOTO 390
380 DECV#=VDEC#+MDEC#/60#
390 DECV#=DECV#*3.14159#/180#:REM CONVERTED INTO RADIANS
400 PRINT
410 INPUT #2,C$
420 PRINT:INPUT #2,CRA,MRAC,SRAC
430 RAC=CRA+MRAC/60+SRAC/3600
440 PRINT:INPUT #2,CDEC,MDECC
450 IF CDEC<>ABS(CDEC) THEN DECC=CDEC+MDECC/60:DECC=-DECC:GOTO 470
460 DECC=CDEC+MDECC/60
470 DECC=DECC*3.14159/180:REM CONVERTED INTO RADIANS
480 PRINT:INPUT #2,N$
490 REM *                          MAIN PROGRAM                            *
500 REM *      RETRIEVES DATA,SEPARATES IT INTO ITS COMPONENTS AND         *
510 REM *      CONVERTS IT TO ARRAYS OF TIME, DEFLECTION AND CODE          *
511 REM EPSILON,B-V,K PRIME AND COLOR DEPENDENT K'S VALUES ARE
512 REM OBTAINED FROM THE INPUT FILE ACCORDING TO THE FORMAT ABOVE
520 INPUT #2,EPSV$:EPSV=VAL(EPSV$)
530 INPUT #2,EPSB$:EPSB=VAL(EPSB$)
540 INPUT #2,BVVAR$:BVVAR=VAL(BVVAR$)
550 INPUT #2,BVCOMP$:BVCOMP=VAL(BVCOMP$)
560 INPUT #2,KPRIV$:KPRIV=VAL(KPRIV$)
570 INPUT #2,KPRIB$:KPRIB=VAL(KPRIB$)
580 INPUT #2,KSECV$:KSECV=VAL(KSECV$)
590 INPUT #2,KSECB$:KSECB=VAL(KSECB$)
600 INPUT #2,REDINIT#
610 INPUT #2,PERIOD#
620 PRINT:PRINT:PRINT
630 PRINT "OBSERVER: ";N$;"   DATE: ";M#;" ";DAY#;" ";YEAR#
635 PRINT "VARIABLE STAR                                    COMPARISON STAR"
640 PRINT V$;"                                                       ";C$
644 PRINT "RA                DEC                               RA               DEC"
648 PRINT " HH  MM   SS.S     DD  MM.M                HH   MM   SS.S      DD   MM.M
650 PRINT VRA#;" ";MRA#;" ";SRA#;"   ";VDEC#;" ";MDEC#;"             ";CRA;" ";MRAC;
" ";SRAC;" ";"   ";CDEC;" ";MDECC
660 PRINT "UBV STANDARDIZATION COEFFICIENTS"
670 PRINT "      EPSILON(V): ";EPSV
680 PRINT "      EPSILON(B): ";EPSB
690 PRINT "B-V VALUES"
700 PRINT "      B-V(VARIABLE): ";BVVAR
710 PRINT "      B-V(COMPARISON: ";BVCOMP
```

```
720 PRINT "EXTINCTION COEFFICIENT VALUES"
730 PRINT "      KPRIME(V): ";KPRIV;"     KSECONDARY(V): ";KSECV
740 PRINT "      KPRIME(B): ";KPRIB;"     KSECONDARY(B): ";KSECB
750 PRINT "EPHEMERIS"
760 PRINT "      TIME OF MIN.: ";REDINIT#;"     PERIOD: ";PERIOD#
770 N=1
780 INPUT #2,DAT$
790 IF MID$(DAT$,1,6)<>"SYSTEM" THEN GOTO 780
800 INPUT #2,DAT$
810 IF DAT$="STOP" THEN PRINT:INPUT"ALL DATA LOADED HIT RETURN TO CONTINUE";DAT$:
KEY ON:CLOSE #2:CLS:GOSUB 3690:GOTO 1290
820 FIL=VAL(MID$(DAT$,1,1))
830 FOR X=1 TO N:IF NOTFIL(X)=FIL THEN 800
840 NEXT X:PRINT N;":filter # is";FIL
850 PRINT "IS THIS FILTER EQUIVALENT TO : "
860 PRINT "      1)BLUE"
870 PRINT "      2)VISUAL"
880 PRINT "      3)NEITHER"
890 INPUT CLOR
900 PRINT:INPUT"DO YOU WANT TO REDUCE THIS FILTER(y/n)";V$
910 IF V$="n" OR V$="N" THEN PRINT V$:NOTFIL(N)=FIL:N=N+1:GOTO 800
920 PRINT V$
930 GOTO 950
940 INPUT #2,DAT$:IF DAT$="STOP" THEN PRINT:INPUT"all data loaded hit return to c
ontinue";DAT$:KEY ON:CLOSE #2:CLS:GOSUB 3690:GOTO 1290
950 NEWFIL=VAL(MID$(DAT$,1,1)):IF (NEWFIL<>FIL) THEN GOTO 940
960 DAT$=" "+ DAT$
970 GOSUB 990
980 GOTO 940
990 REM
1000 KEY OFF
1010 IF MID$(DAT$,2,4)="STOP" THEN GOSUB 1280:GOTO 1420
1020 DAT$=MID$(DAT$,2,16)
1030 FIL=VAL(MID$(DAT$,1,1))
1040 CODE=VAL(MID$(DAT$,2,1))
1050 IF FIL=0 OR MID$(DAT$,2,1)="0" THEN CLS:LOCATE 25,1:PRINT DAT$,"CODING ERROR
";:LOCATE 25,17:BEEP:RETURN
1060 COUNT(FIL)=COUNT(FIL)+1
1070 T$=MID$(DAT$,11,6)
1080 HR=VAL(MID$(T$,1,2))
1090 IF HR<10 THEN HR=HR+24
1100 MIN=VAL(MID$(T$,3,2))/60
1110 SEC=VAL(MID$(T$,5,2))/3600
1120 TIME(FIL,COUNT(FIL))=HR+MIN+SEC
1130 D=VAL(MID$(DAT$,6,4))
1135 DEFL(FIL,COUNT(FIL))=D
1140 REM DEFL(FIL,COUNT(FIL))=(D-204000!)/2040
1150 IF DEFL(FIL,COUNT(FIL))<0 THEN DEFL(FIL,COUNT(FIL))=0
1160 IF DEFL(FIL,COUNT(FIL))>1000 THEN DEFL(FIL,COUNT(FIL))=1000
1170 IF LASTFIL<>FIL THEN CLS:GOSUB 1210
1180  IF LASTFIL=FIL THEN XRAW=TIME(FIL,COUNT(FIL)):YRAW=DEFL(FIL,COUNT(FIL)):GOS
UB 1210
1190 IF DEFL(FIL,COUNT(FIL))=1000 THEN PRINT TAB(20)"FULL SCALE DEFLECTION";
1200 LASTFIL=FIL
1210 REM
1220 CODE(COUNT(FIL))=CODE
1230 D$(COUNT(FIL))=DAT$
1240 X(COUNT(FIL))=TIME(FIL,COUNT(FIL))
1250 Y(COUNT(FIL))=DEFL(FIL,COUNT(FIL))
1260 RETURN
1270 ERASE TIME,DEFL
1280 REM
1290 REM
```

```
1300 REM *                          SUBROUTINE                                  *
1310 REM *    THIS ROUTINE CONVERTS EST TO LOCAL SIDEREAL TIME. THIS            *
1320 REM *   PROGRAM DEALS ONLY WITH CALCULATIONS OF UT AT 75D 21M OF           *
1330 REM *                          GREENWHICH                                  *
1340 REM *                ALMANAC FOR OBSERVER'S METHOD                         *
1350 DIM SYD(270),SID(10),JD#(270),HJD#(270),PHASE#(270),VPHASE#(270),VHJD#(270)
1360 CLS
1370 FOR K=1 TO COUNT(FIL)
1380 STD=X(K)
1390 UT1#=STD-3.5
1400 N=INT((275*M#)/9)-2*INT((M#+9)/12)+DAY#-30
1410 T=N+UT1#/24
1420 X=(T-W)/A-1
1430 FOR I=1 TO 10
1440 SID(I)=0
1450 NEXT I
1460 FOR J=8 TO 1 STEP -1
1470 SID(J)=2*X*SID(J+1)-SID(J+2)+AR(J)
1480 NEXT J
1490 TIME=(SID(1)-SID(3))/2
1500 SDF=TIME+UT1#
1510 SDT=SDF-LONG
1520 IF SDT>24 THEN SDT=SDT-24
1530 IF SDT>24 THEN SDT=SDT-24
1540 SYD(K)=SDT
1550 GOSUB 3410
1560 NEXT K
1570 ERASE AR,SID
1580 PRINT:PRINT:PRINT
1590 REM *                          SUBROUTINE                                  *
1600 REM *   THIS ROUTINE INTERPOLATES THE SKY DEFLECTION READING AT A          *
1610 REM *   PARTICULAR TIME AND SUBTRACTS IT FROM THE COMPARISON AND           *
1620 REM *                   VARIABLE DEFLECTION READING                        *
1630 DIM YS(303),SKY(300),T3(300)
1640 Z=0
1650 FOR I=1 TO COUNT(FIL)
1660    J=CODE(I)
1670    IF J=4 THEN GOSUB 3340
1680 NEXT I
1690 PRINT"X SUN:  ";X#;"      Y SUN:    ";Y#
1700 PRINT:PRINT:PRINT:
1710 Z=0
1720 PRINT"DATA POINT        TIME        DEFL       DEFL-SKY  JD(GEO)         CODE NU
M"
1730 FOR I=1 TO COUNT(FIL)
1740 IF CODE(I)=4 THEN GOSUB 3380:YS(I)=0:GOTO 1770
1750 SUBB=SKY(Z)+((X(I)-T3(Z))/(T3(Z+1)-T3(Z)))*(SKY(Z+1)-SKY(Z))
1760 YS(I)=Y(I)-SUBB
1770 IF X(I)>=24 THEN PRINT USING "&   ##.#####  ###.####   ###.####   #######.#####
  #     ###";D$(I),X(I)-24,Y(I),YS(I),JD#(I),CODE(I),I ELSE PRINT USING "&   ##.###
##  ###.####   ###.####   #######.#####   #      ###";D$(I),X(I),Y(I),YS(I),JD#(I),COD
E(I),I
1780 NEXT I
1790 ERASE D$,Y
1800 REM *                        HOUR ANGLE                                    *
1810 DIM HA(303),HAC(300)
1820 FOR I=1 TO COUNT(FIL)
1830 J=CODE(I)
1840 IF J=3 THEN HA(I)=(SYD(I)-RAC)*15*3.14159/180
1850 IF J=1 THEN HA(I)=(SYD(I)-RAV#)*15*3.14159/180:HAC(I)=(SYD(I)-RAC)*15*3.1415
9/180
1860 NEXT I
```

```
1870 PRINT:PRINT:PRINT
1880 PRINT "SECZ"," MAG"
1890 REM *               CALCULATE SEC Z AND APPARENT MAGNITUDES              *
1900 DIM SECZC(300),MC(300),SECZV(300),MV(300),CSECZ(300),NC(300)
1910 DIM T2(200),T1(200)
1920 FOR I=1 TO COUNT(FIL)
1930 J=CODE(I)
1940 IF J=3 THEN GOSUB 3110
1950 IF J=1 THEN GOSUB 3210
1960 NEXT I
1970 ENDCOM=Q
1980 ENDVAR=R
1990 REM
2000 REM *                        LEAST SQUARES                          *
2010 DIM DMAG(200),DELMAG(200),XBAR(200)
2020 FOR I=1 TO ENDCOM
2030 SUMXY=SUMXY+SECZC(I)*MC(I)
2040 SUMXSQ=SUMXSQ+(SECZC(I)^2)
2050 SUMX=SUMX+SECZC(I)
2060 SUMY=SUMY+MC(I)
2070 NEXT I
2080 M=(ENDCOM*(SUMXY)-SUMX*SUMY)/(ENDCOM*(SUMXSQ)-(SUMX)^2)
2090 B=(SUMY-M*SUMX)/ENDCOM
2100 K=M
2110 FOR I=1 TO ENDCOM
2120 Y1=M*SECZC(I)+B
2130 Y2=MC(I)
2140 DELTA=Y2-Y1 : DELSQ=DELTA^2
2150 DELQS=DELSQ^.5
2160 EOI=EOI+DELQS
2170 NEXT I
2180 FOR I=1 TO F
2190 IF NC(I)=3 THEN H=H+1
2200 IF NC(I)=1 THEN GOSUB 3060
2210 NEXT I
2220 FOR I=1 TO ENDCOM
2230 PRINT SECZC(I),MC(I)
2240 NEXT I
2250 PRINT "THE CALCULATED EXTINCTION COEFFICIENT IS ";K
2260 PRINT "DO YOU WISH TO :"
2270 PRINT "     1)INPUT YOUR OWN EXTINCTION COEFFICIENT FOR THIS FILTER?"
2280 PRINT "     2)CALCULATE THE EXTINCTION COEFFICIENT BASED ON COLOR"
2290 PRINT "       DIFFERENCES BETWEEN COMP AND VARIABLE STAR?"
2300 PRINT "     3)KEEP THE CALCULATED VALUE":INPUT F:IF F=3 GOTO 2330
2310 IF F=1 THEN INPUT"INPUT YOUR VALUE";K:PRINT "THE EXTINCTION COEFF.: ";K:GOTO
 2340
2320 IF CLOR=2 THEN KV=KPRIV+KSECV*BVVAR AND KC=KPRIV+KSECV*BVCOMP ELSE KV=KPRIB+
KSECB*BVVAR AND KC=KPRIB+KSECB*BVCOMP:PRINT "THE EXTINCTION COEFF. FOR THE VARIAB
LE STAR: ";KV;"  FOR THE COMP STAR: ";KC
2330 PRINT
2340 REM
2350 PRINT "THE PROBABLE ERROR IS: ";EOI/ENDCOM
2360 PRINT:PRINT:PRINT
2370 PRINT"TIME         JD(HEL)           PHASE     DMAG"
2380 FOR I=1 TO ENDVAR
2390 GOSUB 4480
2400 IF T1(I)>=24 THEN PRINT USING"###.######  #######.#####  ##.#####  ##.#####"
;T1(I)-24,VHJD#(I),VPHASE#(I),DMAG(I) ELSE PRINT USING"###.######  #######.#####
 ##.#####  ##.#####";T1(I),VHJD#(I),VPHASE#(I),DMAG(I)
2410 NEXT I
2420 REM *                PLOTTING ROUTINE: DMAG VS PHASE                 *
2430 REM INITIAL VALUES ARE SET TO THE FIRST VALUE OF EACH ARRAY
2440 DMAX=ABS(DMAG(1)):DMIN=ABS(DMAG(1)):PHASEMAX=VPHASE#(1):PHASEMIN=VPHASE#(1)
```

```
2450 REM THIS LOOP CHECKS THE VALUES IN THE ARRAYS WITH THE
2460 REM MAX AND MIN VALUES AND ULTIMATELY DETERMINES THE TRUE
2470 REM VALUES FOR THE MAX AND MIN OF EACH ARRAY
2480 LABL$(1)="DMAG VS PHASE"
2490 LABL$(2)="PHASE"
2500 LABL$(3)="DMAG"
2510 PHASEMA=(INT(VPHASE#(ENDVAR)*1000))/1000:PHASEMI=(INT(VPHASE#(1)*1000))/1000
2520 FOR I=1 TO ENDVAR
2530 IF VPHASE#(I)<VPHASE#(I+1) THEN GOTO 2550
2540 VPHASE#(I)=VPHASE#(I)+1:I=I+1:IF I>ENDVAR GOTO 2560 ELSE GOTO 2540
2550 NEXT I
2560 FOR I=1 TO ENDVAR
2570 IF ABS(DMAG(I))>DMAX THEN DMAX=ABS(DMAG(I))
2580 IF ABS(DMAG(I))<DMIN THEN DMIN=ABS(DMAG(I))
2590 IF VPHASE#(I)>PHASEMAX THEN PHASEMAX=VPHASE#(I)
2600 IF VPHASE#(I)<PHASEMIN THEN PHASEMIN=VPHASE#(I)
2610 NEXT I
2620 REM THIS LOOP SUBTRACTS THE MIN VALUE OF EACH ARRAY FROM EVERY
2630 REM VALUE IN THAT ARRAY
2640 FOR I=1 TO ENDVAR
2650 VPHASE#(I)=VPHASE#(I)-PHASEMIN
2660 DMAG(I)=ABS(DMAG(I))-DMIN
2670 NEXT I
2680 REM THIS LOOP DIVIDES EVERY NUMBER BY THE MAX VALUE IN THAT
2690 REM PARTICULAR ARRAY AND THEN MULTIPLIES THAT RESULT BY
2700 REM 1000 THUS NORMALIZING THE ARRAY TO A 0 TO 1000 SCALE
2710 FOR I=1 TO ENDVAR
2720 DMAG(I)=DMAG(I)/DMAX*1000
2730 VPHASE#(I)=VPHASE#(I)/PHASEMAX*1000
2740 NEXT I
2750 REM THIS IS THE ACTUAL PLOTTING OF THE POINTS ROUTINE
2760 REM FIRST AXES ARE DRAWN THEN THE SCALED POINTS ARE PLOTTED
2770 REM (DMAG VS PHASE)
2780 KEY OFF
2790 CLS:SCREEN 1
2800 XS=530/1000
2810 YS=180/1000
2820 LINE(10,0)-(10,180)
2830 LINE(10,180)-(639,180)
2840 FOR I=1 TO ENDVAR
2850 C=VPHASE#(I)*XS+10:D=180-DMAG(I)*YS
2860 C=INT(C):D=INT(D)
2870 GOSUB 4260
2880 NEXT I
2890 L=LEN(LABL$(2)):P=INT((80-P)/2)
2900 LOCATE 24,P:PRINT LABL$(2);
2910 L=LEN(LABL$(3)):P=INT((25-L)/2)
2920 FOR I=1 TO L
2930 LOCATE P,1
2940 PRINT MID$(LABL$(3),I,1)
2950 P=P+1:NEXT I
2960 LOCATE 3,20:PRINT LABL$(1)
2970 LOCATE 25,1
2980 PRINT "X RANGE :";PHASEMI;" TO ";PHASEMA
2990 PRINT "Y RANGE :-";DMIN;" TO -";DMAX
3000 LOCATE 24,1
3010 T$=INKEY$:IF T$="" THEN 3010
3020 SCREEN 0:WIDTH 80:CLS
3030 END
3040 REM THIS ROUTINE CALCULATES THE DELTA MAGNITUDE FOR EXTINCTION
3050 REM COEFFICIENT CALCULATION
```

```
3060 L=L+1
3070 FAFID=((T1(L)-T2(H))/(T2(H+1)-T2(H)))
3080 DELMAG(L)=MV(L)-(MC(H)+(FAFID*(MC(H+1)-MC(H)))):XBAR(L)=.5*(MV(L)+(MC(H)+(FA
FID*(MC(H)))))
3090 RETURN
3100 REM THIS ROUTINE CALCULATES SEC Z AND MAG FOR THE COMPARISON STAR
3110 Q=Q+1
3120 F=F+1
3130 COSZ=COS(LAT-DECC)-2*COS(DECC)*COS(LAT)*(SIN(HA(I)/2)^2)
3140 REM AIRMASS APPROXIMATION FROM YOUNG(METHODS OF EXPERIMENTAL PHYSICS
3150 REM                          VOL 12(1974)
3160 SECZC(Q)=1/COSZ:SECZC(Q)=SECZC(Q)*(1-.0012*(SECZC(Q)-1)):NC(F)=3
3170 MC(Q)=(-2.5/LOG(10))*LOG(YS(I)/1000):T2(Q)=X(I)
3180 RETURN
3190 REM THIS ROUTINE CALCULATES SEC Z AND MAG FOR THE VARIABLE STAR
3200 REM AND ALSO THE FOR THE COMPARISON STAR AT THIS PARTICULAR TIME
3210 R=R+1
3220 F=F+1
3230 COSZ=COS(LAT-DECV#)-2*COS(DECV#)*COS(LAT)*(SIN(HA(I)/2)^2)
3240 REM AIRMASS APPROXIMATION FROM YOUNG(METHODS OF EXPERIMENTAL PHYSICS
3250 REM                          VOL 12(1974)
3260 SECZV(R)=1/COSZ:SECZV(R)=SECZV(R)*(1-.0012*(SECZV(R)-1)):NC(F)=1
3270 MV(R)=(-2.5/LOG(10))*LOG(YS(I)/1000):T1(R)=X(I)
3280 CCOSZ=COS(LAT-DECC)-2*COS(DECC)*COS(LAT)*(SIN(HAC(I)/2)^2)
3290 REM AIRMASS APPROXIMATION FROM YOUNG(METHODS OF EXPERIMENTAL PHYSICS
3300 REM                          VOL 12(1974)
3310 CSECZ(R)=1/CCOSZ:CSECZ(R)=CSECZ(R)*(1-.0012*(CSECZ(R)-1))
3320 RETURN
3330 REM THIS ROUTINE IS CALLED FROM THE ROUTINE FOR SUBTRACTION OF SKY
3340 Z=Z+1
3350 SKY(Z)=Y(I)
3360 T3(Z)=X(I)
3370 RETURN
3380 Z=Z+1
3390 RETURN
3400 REM THIS ROUTINE CALCULATES JULIAN DAY(GEOCENTRIC AND HELIOCENTRIC)
3410 JD#(K)=367#*YEAR#-INT((7#*(YEAR#+INT(((M#+9#)/12#)))/4#))+INT((275#*M#/9#))+
DAY#+1721013.5#+UT1#/24#
3420 RAV#=RAV#*15#*3.14159#/180#:REM converted into degrees then radians
3430 TARGU#=JD#(K)-2451545#
3440 G#=357.528#+.9856003#*TARGU#
3450 IF G#<360# THEN G#=G#+360#:GOTO 3450
3460 G#=G#*3.14159#/180#
3470 L#=280.46#+.9856474#*TARGU#
3480 IF L#<360# THEN L#=L#+360#:GOTO 3480
3490 LAMBDA#=L#+1.915#*SIN(G#)+.02#*SIN(2*G#)
3500 LAMBDA#=LAMBDA#*3.14159#/180#
3510 EPSILON#=23.439#-.0000004#*TARGU#
3520 EPSILON#=EPSILON#*3.14159#/180#
3530 ARE#=1.00014#-.01671#*COS(G#)-.00014#*COS(2*G#)
3540 X#=ARE#*COS(LAMBDA#)
3550 Y#=ARE#*COS(EPSILON#)*SIN(LAMBDA#)
3560 DELT#=-.00577#*((COS(DECV#)*COS(RAV#))*X#+(SIN(DECV#)*TAN(EPSILON#)+COS(DECV
#)*SIN(RAV#))*Y#)
3570 HJD#(K)=JD#(K)+DELT#:IF CODE(K)=1 THEN VHJD#(U)=HJD#(K):U=U+1
3580 RAV#=RAV#*180#/(3.14159#*15#)
3590 GOSUB 3610
3600 RETURN
3610 REM *          PHASE CALCULATION(WRITTEN BY K.GARLOW)             *
3620 REM *   THE INITIAL JD VALUE AND PERIOD VALUE ARE SPECIFIC TO     *
3630 REM *     THE W UMA BINARY SYSTEM AND WILL HAVE TO BE CHANGED     *
3640 REM *                        AS NEEDED                            *
```

```
3650 DELDAYS#=HJD#(K)-REDINIT#
3660 MODULO#=DELDAYS#-INT(DELDAYS#/PERIOD#)*PERIOD#
3670 PHASE#(K)=MODULO#/PERIOD#:IF CODE(K)=1 THEN VPHASE#(V)=PHASE#(K):V=V+1
3680 RETURN
3690 REM *               PLOTTING ROUTINE FOR DEFLECTION              *
3700 REM *                            VS. EST                          *
3710 INPUT "DO YOU WISH TO PLOT DEFLECTION VS. EST FOR THIS FILTER ";H$
3720 IF H$="NO" OR H$="N" THEN RETURN
3730 LABEL$(1)="DEFL VS. TIME"
3740 LABEL$(2)="TIME"
3750 LABEL$(3)="DEFL"
3760 LABEL$(4)="0":LABEL$(5)="1000"
3770 TEMP=X(1):DIFF=X(COUNT(FIL))-TEMP
3780 FOR I=1 TO COUNT(FIL)
3790 X(I)=X(I)-TEMP:X(I)=X(I)/DIFF*1000
3800 NEXT I
3810 INPUT "ENTER 1 FOR HIRES OR 2 FOR MEDRES,C/R TO ABORT";T$
3820 IF T$="1" THEN XS=630/1000:YS=180/1000 :SCREEN 2: GOTO 3850
3830 IF T$="2" THEN XS=310/1000:YS=180/1000 :SCREEN 1:GOTO 3850
3840 RETURN
3850 CLS
3860 LINE (10,0)-(10,180)
3870 LINE (10,180)-(639,180)
3880 FOR I=1 TO COUNT(FIL)
3890 X=X(I)*XS+10
3900 Y=180-Y(I)*YS
3910 X=INT(X):Y=INT(Y)
3920 GOSUB 4300
3930 NEXT I
3940 FOR I=1 TO 7
3950 P=20
3960 LABEL$(I)=MID$(LABEL$(I),1,P)
3970 NEXT I
3980 L=LEN(LABEL$(2))
3990 P=INT((80-P)/2)
4000 LOCATE 24,P
4010 PRINT LABEL$(2);
4020 L=LEN(LABEL$(3))
4030 P=INT((25-L)/2)
4040 FOR I=1 TO L
4050 LOCATE P,1
4060 PRINT MID$(LABEL$(3),I,1)
4070 P=P+1
4080 NEXT I
4090 IF T$="1" THEN LOCATE 2,60
4100 IF T$="2" THEN LOCATE 3,20
4110 PRINT LABEL$(1)
4120 KEY OFF
4130 LOCATE 25,1
4140 X$="Y RANGE :"+LABEL$(4)+ " TO "+LABEL$(5)
4150 PRINT X$;
4155 IF TEMP>24 THEN TEMP=TEMP-24
4156 HGH=DIFF+TEMP:IF HGH>24 THEN HGH=HGH-24
4160 PRINT TAB(40) "X RANGE :";TEMP;" TO ";HGH
4170 LOCATE 24,1
4180 T$=INKEY$:IF T$="" THEN 4180
4190 SCREEN 0:WIDTH 80:CLS
4200 KEY ON:COLOR 14,1,4:CLS
4210 X=0:Y=0:P=0:L=0:XS=0:YS=0
4220 FOR I=1 TO COUNT(FIL)
```

```
4230 X(I)=X(I)/1000*DIFF:X(I)=X(I)+TEMP
4240 NEXT I
4250 RETURN
4260 REM
4270 SHAPE$(1)="BM"+STR$(C)+","+STR$(D)+"NE1NF1NG1NH1"
4280 DRAW SHAPE$(1)
4290 RETURN
4300 REM
4310 SHAPE$(1)="BM"+STR$(X)+","+STR$(Y)+"NE1NF1NG1NH1"
4320 SHAPE$(2)="BM"+STR$(X)+","+STR$(Y)+"NU1NR1ND1NL1"
4330 SHAPE$(3)="BM"+STR$(X)+","+STR$(Y)+"NU1NR1ND1NL1"
4340 SHAPE$(4)="BM"+STR$(X)+","+STR$(Y)+"M+0,+0"
4350 DRAW SHAPE$(CODE(I))
4360 RETURN
4365 REM SIDEREAL TIME CALCULATION CONSTANTS FROM ALMANAC FOR OBSERVER'S
4366 REM THESE CONSTANTS ARE ACCURATE FOR THE YEAR 1985 AFTER WHICH THEY
4367 REM                     MUST BE ADJUSTED
4370 AR(1)=37.4299232#
4380 AR(2)=12.02493092#
4390 AR(3)=4.01E-06
4400 AR(4)=9.4E-07
4410 AR(5)=7.45E-06
4420 AR(6)=1.625E-05
4430 AR(7)=-4.66E-06
4440 AR(8)=-.0000067
4450 A=183
4460 W=1
4470 RETURN
4480 REM DELTA MAGNITUDE CALCULATION FOR B,V AND DEFAULT METHOD
4485 IF CLOR=1 THEN DMAG(I)=DELMAG(I)-KPRIV*(SECZV(I)-CSECZ(I))-KSECV*XBAR(I)*(BV
VAR-BVCOMP)+EPSV*(BVVAR-BVCOMP):RETURN
4490 IF CLOR=2 THEN DMAG(I)=DELMAG(I)-KPRIB*(SECZV(I)-CSECZ(I))-KSECB*XBAR(I)*(BV
VAR-BVCOMP)+EPSB*(BVVAR-BVCOMP):RETURN
4500 DMAG(I)=DELMAG(I)+K*(CSECZ(I)-SECZV(I)):RETURN
```

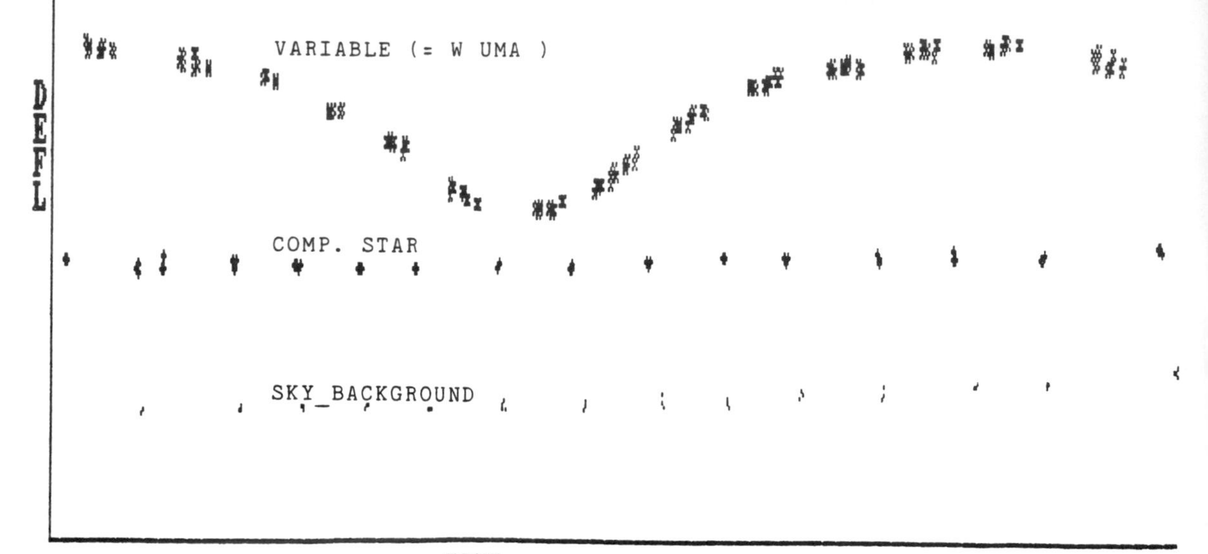
DEFL VS. TIME
SAMPLE OUTPUT PLOT FROM COMPUTER REDUCTION PROGRAM: PLOT OF INPUT DATA OF VARIABLE STAR, COMPARISON STAR, AND SKY.
VARIABLE (= W UMA)
DEFL
COMP. STAR
SKY_BACKGROUND
TIME
Y RANGE :0 TO 1000
X RANGE : 22.55583 TO 27.48361

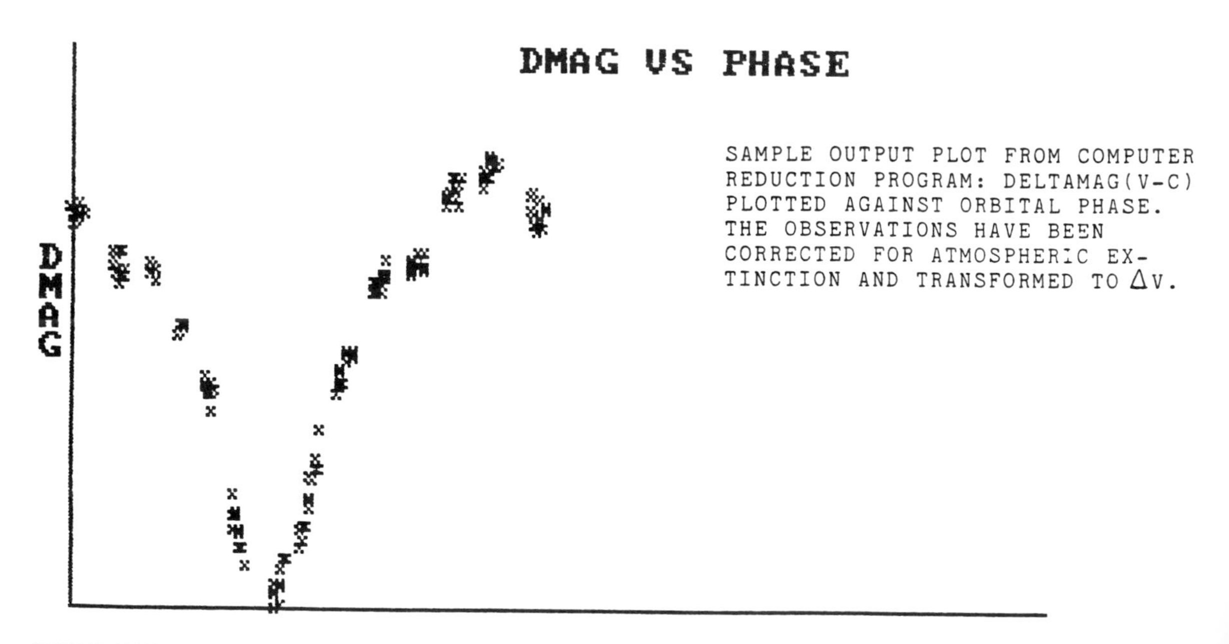
DMAG VS PHASE
SAMPLE OUTPUT PLOT FROM COMPUTER REDUCTION PROGRAM: DELTAMAG(V-C) PLOTTED AGAINST ORBITAL PHASE. THE OBSERVATIONS HAVE BEEN CORRECTED FOR ATMOSPHERIC EX-TINCTION AND TRANSFORMED TO ΔV.
DMAG
PHASE

P CYGNI: A HYPERGIANT WELL WORTH OBSERVING

Mart de Groot
Armagh Observatory
College Hill
Armagh, BT61 9DG
Northern Ireland

Angelo Cassatella
Astronomy Division, ESTEC
Villafranca Satellite Tracking Station
European Space Agency
Apartado 54065
Madrid, Spain

Henny J.G.L.M. Lamers
Astronomical Institute
Space Research Laboratory
Beneluxlaan 21
3527 HS Utrecht
The Netherlands

I. INTRODUCTION

P Cygni (=34 Cyg = HR 7763 = HD 193237 = BD +37°3871) is a hypergiant (an especially luminous supergiant) of spectral type B1 Ia$^+$. It was discovered in the year 1600 when it rose in brightness and reached third magnitude. After rather drastic brightness variations in the 17th century the star settled at $V = 4.9$, showing only small variations with $\Delta V \approx 0.2$ mag (Schneller 1957; de Groot 1969). The star has long been considered as the prototype object of the so-called *P Cygni-type stars*. While originally this class of stars included rather diverse objects (see *e.g.*, de Groot 1973) the class has recently been redefined by Lamers who proposed the following definintion of P Cygni-type (PCT) stars: "P Cygni-type stars are luminous supergiants (with $M_V \lesssim -7$) of spectral types O, B, and A, which in their visual spectrum show or have shown P Cygni profiles (emission lines with blue-shifted absorption components) of the strongest Balmer and He I lines (Hα, Hβ, HeI 6678, HeI 5875)" (Lamers 1985).

As defined above, PCT stars are located in the upper part of the Hertzsprung-Russell diagram (HRD) close to the empirical upper luminosity limit (Humphreys & Davidson 1979) now called the Humphreys-Davidson limit (Van Gent & Lamers 1985). PCT stars are believed to represent a relatively short-lived phase of the evolution of massive stars (Lamers 1985), whose close investigation can contribute greatly to our understanding of the physical mechanisms playing a crucial role in this part of the HRD.

This contribution presents the scientific reasons for making regular photometric observations of the prototype star of the class, P Cygni.

II. P CYGNI'S BASIC PARAMETERS

P Cygni's basic parameters were determined by Lamers *et al.* (1983) who give the following quantities: Sp. Type: B1 Ia$^+$; $E(B-V) = 0.63 \pm 0.05$; distance $= 1.8 \pm 0.1$ kpc; $T_{eff} = 19300 \pm 2000$ K; $R_* = 76 \pm 14 R_\odot$; $\log L/L_\odot = 5.86 \pm 0.30$; $M_V = -8.3 \pm 0.2$; $M_{bol} = -9.9 \pm 0.7$.

With these parameters, it is possible to plot P Cygni on the HRD (see Figure 1). Figure 1 shows the upper part of the main sequence with evolutionary tracks (including mass loss) for stars with initial masses of 60 and 85 $M_\odot$. These tracks are from Maeder (1983) for his higher mass-loss rates, as appropriate for P Cygni. The areas in the HRD where one finds red supergiants (RSG), late WN stars (a sub-class of Wolf-Rayet stars), and the so-called S Dor variables are also marked. The Humphreys-Davidson limit is also indicated. In Figure 1 we also indicate the positions of some other PCT and S Dor type stars; R71 and R81 are S Dor variables in the Large Magellanic Cloud (LMC) (Van Genderen 1971a, b; Wolf *et al.* 1981a, b); ς^1 Sco is a galactic PCT star (Appenzeller & Wolf 1979; Hutchings 1979); Cyg OB2 No. 12 is a galactic B-type hypergiant hidden behind about ten magnitudes of interstellar absorption (Morgan *et al.* 1954; Humphreys 1978); η Car is a unique object, but with certain similarities to the other stars marked.

It is especially interesting to note that the evolution, with mass loss, of massive stars undergoes a rather sudden change when passing a critical initial mass of about $50 M_\odot$. Below this critical mass, stars which have left the main sequence after the completion of hydrogen core burning will cross the HRD and become RSG's before returning to the left-hand side of the HRD. Above the critical mass, stars will not move over as far as the RSG phase but will return to the left part of the HRD after a redward loop which is shorter for stars of higher initial mass. This effect is caused primarily by heavy mass loss, and should be considered one of the principal arguments to explain the observed Humphreys-Davidson limit.

From P Cygni's location in the HRD, two facts are apparent: a) P Cygni is in the borderline area below which a star's evolution brings it through a RSG phase and above which the RSG phase is absent, and b) P Cygni is in the same general area as the S Dor variables, although very near their blue limit. Thus, we

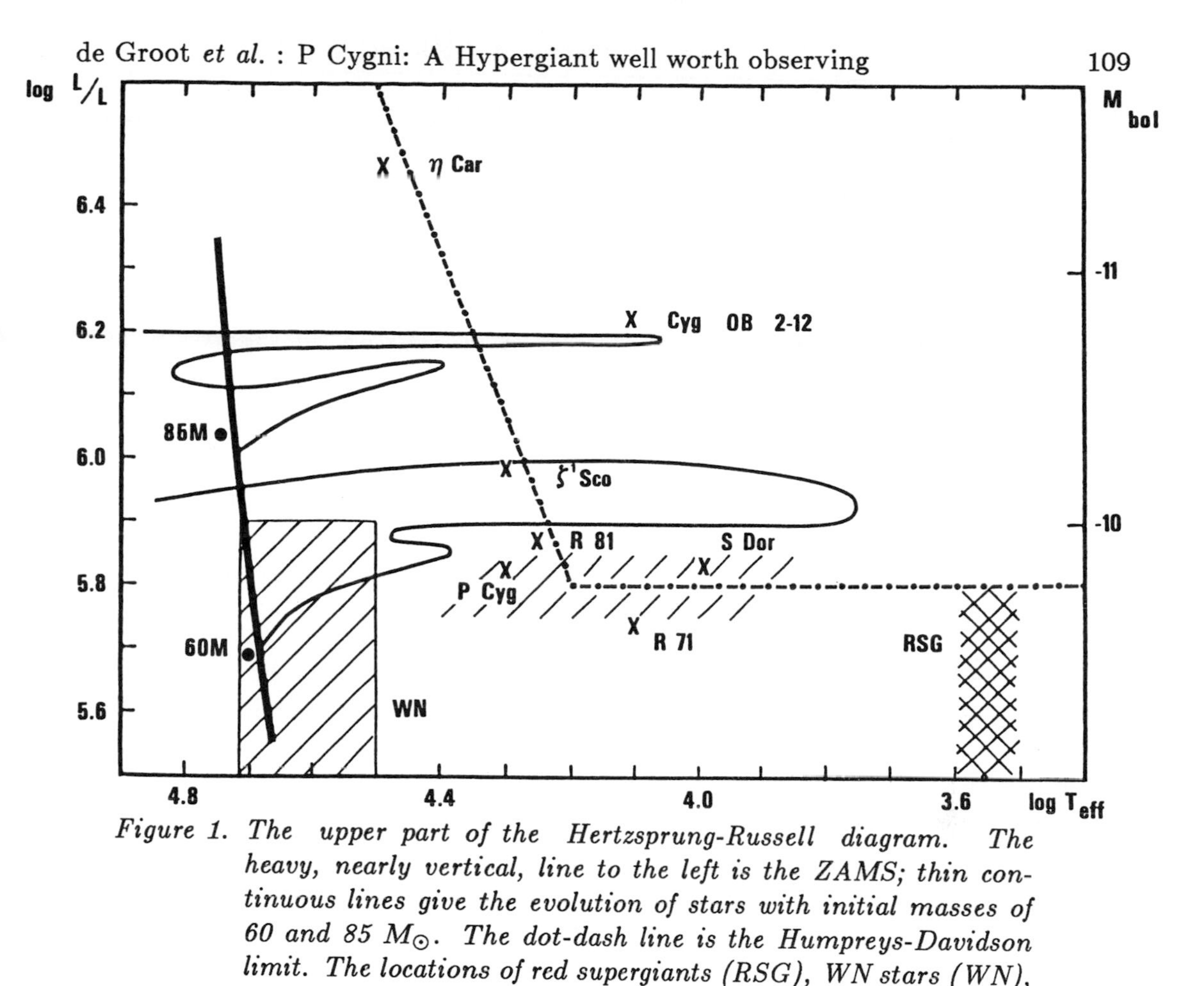

Figure 1. The upper part of the Hertzsprung-Russell diagram. The heavy, nearly vertical, line to the left is the ZAMS; thin continuous lines give the evolution of stars with initial masses of 60 and 85 $M_\odot$. The dot-dash line is the Humpreys-Davidson limit. The locations of red supergiants (RSG), WN stars (WN), and S Dor variables are indicated, as well as the positions of a few selected objects. (Adapted from Maeder 1983.)

are prompted to ask the following questions: i) Has P Cygni ever gone, or will it ever go, through a RSG phase, and ii) What is P Cygni's relation to the S Dor variables?

Before we try to answer these questions let us summarize the relevant observations of P Cygni.

III. OBSERVED VARIATIONS OF P CYGNI

P Cygni's photometric history has been documented by Schneller (1957) and has also been given by, among others, de Groot (1969) and Lamers (1985). A somewhat schematic representation of the light curve is given in Figure 2. The

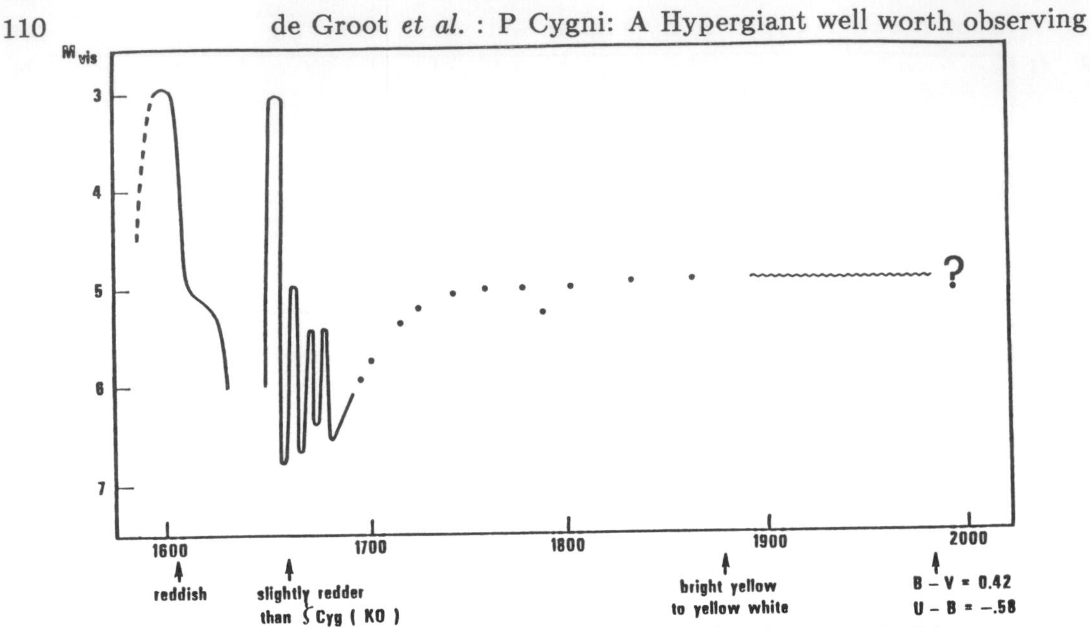

Figure 2. A schematic representation of P Cygni's photometric history from 1600-1985. The colour is given below the abscissa.

star's colour has been indicated below the abscissa. Note that it was red in the 17th century, slowly becoming bluer since.

The photometric variations have been studied by many authors. Pigott (1786) postulated an 18-year period. From observations in the 1930's, Kharadze *et al.* (1936) and Nikonov (1937, 1938) found indications for a 27-day period. Observations in the 1950's led Kharadze & Magalashvili (1967) and Magalashvili & Kharadze (1967a, b) to announce a period of $0.^{d}500659$ with a W UMa type light curve. When this was checked by Alexander & Wallerstein (1967), no short-period variations were found. Fernie (1968, 1969) and Percy & Welch (1983) found evidence for variations on a time scale between 30 and 50 days. More recent observations by Smeets *et al.* (1985) revealed a 60-day time-scale not strictly periodic.

Spectroscopic variations have also been observed. The Balmer lines of hydrogen often show multiple absorption components, and the one at highest blueward displacement was found by de Groot (1969) to vary its radial velocity with a 114-day period during the interval 1942-1964. A similar study by Luud *et al.* (1975) showed a 57-day period for the main Balmer absorption component at lower displacements during 1965-1971. Markova & Kolka (1984) observed similar spectral variations in 1981, but interpreted these as the ejection of successive shells about every two months. A similar interpretation was given to faint secondary components observed in the ultraviolet by Lamers *et al.* (1984, 1985), albeit at a time scale of one shell every 1-2 years. Finally, polarimetric observations by Hayes in 1978-79 (Hayes 1985) indicated the presence of 12 and 125-day time-scales.

There is thus a wide variety of time-scales for photometric and spectroscopic variations of P Cygni. Add to this a rotational period less than 72 days (derived from the observed $v \sin i = 75$ km s^{-1} and the above value of R_*) and the confusion is as complete as can be. Recently, Van Gent & Lamers (1985) have re-analyzed all the above observations. They find that many of the time-scales and periodicities claimed do not stand up to Fourier type significance tests and that the best established time-scale for episodic mass loss is one of about 25 days. This fits in nicely with the period for non-radial pulsations $P_{puls} = Q(\bar{\rho}/\bar{\rho}_\odot)^{1/2} \approx 28$ days for P Cygni (Van Gent & Lamers 1985).

IV. VISUAL EVIDENCE OF MASS LOSS

There is one further observation which bears on the questions asked at the end of section II. Wendker (1982) has reported the presence, at a wavelength of 6 cm, of a radio arc which looks like a bow shock with P Cygni at its head. The length of the arc is approximately 75″, which corresponds to a linear distance of 0.65 pc. If, as Wendker suggests, the arc is the result of the outburst in AD 1600, then an expansion velocity of 800 km s^{-1} is required. While this is modest for a normal early B-type supergiant, it is far too high a velocity for P Cygni which today shows a terminal velocity in its wind of about 300 km s^{-1}.

Another possibility, also suggested by Wendker, is that P Cygni has a space velocity which is high compared to the ejection velocity, so that it is seen at the head of the bow shock. If we now assume that the matter in the arc was ejected from the star while it was a RSG or while it evolved from the RSG phase back to the left-hand side of the HRD, then, at a wind velocity of 25 km s^{-1}, which is normal for a RSG, the farthest visible parts of the arc were ejected about 3×10^4 years ago. This time interval is comparable to the time necessary for a star of initial mass 60 $M_\odot$ to evolve from a RSG to an early B type star: 4×10^4 yrs according to Maeder (1981). Thus, the radio arc may be a visible remnant of P Cygni's mass loss during the past 3 to 4×10^4yrs.

V. COMPARISON WITH S DOR VARIABLES

As mentioned in section II, P Cygni is in the same general area of the HRD as the S Dor variables. Humphreys & Davidson (1979) explained the light variations and variable mass-loss rates of the S Dor variables as a result of their high general mass-loss rates which would, possibly together with turbulent pressure, lead to episodes of catastrophic mass loss which make a star move to the left in the HRD. Thus, stars are prevented from crossing the observed upper luminosity limit and the S Dor variables are the stars nearest the limit and unstable from time to time.

S Dor variables show rather large photometric variations, typically at least one magnitude over a 10-year interval. That P Cygni does not show such variations at the present may mean that it is an S Dor variable in a rather quiescent phase. Its 1600 outburst happened while it was more unstable and probably closer to the Humphreys-Davidson limit. The matter ejected in the 17th century obscured the star, making it both rather dim and red. As the dust has blown away and evaporated, the star became bluer and recovered some of its brightness.

Questions i) and ii) at the end of section II are thus tentatively answered as follows: P Cygni has evolved away from the main sequence and several 10^4 years ago was a RSG. Now it is an early B-type hypergiant in a state of relative quiescence similar to that exhibited by some S Dor variables (*e.g.*, R 66: Stahl *et al.* 1983). Periods of quiescence are alternated with periods of great activity, the last one of which occurred in the 17th century. These active phases contribute to the star's losing enough surface material so that the products of nuclear reactions deeper down may become more fully exposed and the star will become a late WN star (Lamers *et al.* 1983a; Lamers 1985).

VI. WHY PHOTOMETRY IS NEEDED TODAY

From the foregoing it should have become clear that P Cygni is an enigmatic star posing a number of intriguing questions. It is a bright star whose spectral variations are being reasonably well covered, both at visual and ultraviolet wavelengths. This is not to say that more spectroscopic observations would mean a waste of time. On the contrary, especially well-planned regular spectroscopic observations at high dispersion will be very useful.

However, compared to spectroscopy, photometry of P Cygni is not done frequently enough at all. One of the reasons may be its brightness, which makes it an awkward object for most professional instruments. But small telescopes will have no problem here.

Any observer who wants to include P Cygni in his/her program should realize that the most characteristic time scale for photometric variations is of the order of one month. This means that regular observations over a complete observing season are needed. Again, this is a requirement sometimes difficult to fulfill by professional astronomers. Amateur observers with access to a small telescope with a photoelectric photometer are in the ideal position to contribute substantially to our photometric coverage and to find answers to the above questions.

What is needed is differential photometry to an accuracy of about $0.^{m}005$ and at a frequency of at least once a week, preferably every 2-3 days. Photometric systems employing filters with narrower band widths may be more suitable from the astrophysical point of view, but *UBV* photometry seems to be the more feasible from the small-telescope observer's position.

It is recommended that all observers use the same comparison star, 36 Cygni, and the same check star, HR 7757. If desired, HR 7613 can be used as an alternative comparison star. [*Editor's Note*: the latter star is used as the primary comparison star for P Cyg in the Bright Be Star Campaign (see article by Percy elsewhere in this volume).]

Observers are of course free to publish their results in their preferred way, but it would be highly appreciated if the author of this article (MdG) would receive a copy of the results as soon as possible after they have been obtained. Of course, all such contributions will be duly acknowledged in a subsequent publication.

In summary, the main reason for good photometry is to find an answer to the question about the possible correlation between light and spectrum variations. In other words, is P Cygni's episodic mass loss reflected in its light variations? The above-mentioned spectroscopic observations of Markova & Kolka (1984) and the photometric observations of Smeets *et al.* (1985) may contain a partial answer to this question, but more systematic observations are needed before the full answer can be known. Furthermore, in the S Dor variables in the LMC the photometric and spectroscopic variations are quite clearly correlated (Stahl *et al.* 1984; Wolf *et al.* 1981b).

The other interesting question, and one which takes us beyond the more descriptive study of P Cygni into the area of understanding the cause of the variations and maybe even the outbursts, is: Do the observed variations compare well with stellar evolution calculations and other theoretical predictions, *e.g.*, non-radial pulsations?

Finally, bearing in mind the 100-1000 year time-scale for major outbursts of P Cygni suggested by Humphreys & Davidson (1979), and the fact that the last outburst happened more than 300 years ago, it is just possible that P Cygni will have another eruption very soon. Photometry by dedicated observers with small telescopes would be the simplest way to detect such an outburst in its earliest phases, whereupon the spectroscopic observers can be alerted and come into action.

REFERENCES

Alexander, T. & Wallerstein, G. (1967). *Pub. A. S. P.*, **79**, 500.
Appenzeller, I. & Wolf, B. (1979). *Astr. Ap. Suppl.*, **38**, 51.
De Groot, M. (1969). *Bull. Astr. Inst. Neth.*, **20**, 225.
De Groot, M. (1973). In *Wolf-Rayet and High-Temperature Stars*, ed. M.K.V. Bappu & J. Sahade, p. 108. Dordrecht: D. Reidel.
Fernie, J.D. (1968). *The Observatory*, **88**, 167.
Fernie, J.D. (1969). *Pub. A. S. P.*, **81**, 168.
Hayes, D.P. (1985). *Ap. J.*, **289**, 726.
Humphreys, R.M. (1978). *Ap. J. Suppl.*, **38**, 309.
Humphreys, R.M. & Davidson, K. (1979). *Ap. J.*, **232**, 409.
Hutchings, J.B. (1979). *Ap. J.*, **233**, 913.
Kharadze, E.K., Nikonov, V., & Kulikovsky, P. (1936). *The Observatory*, **59**, 88.
Kharadze, E.K. & Magalashvili, N.L. (1967). *The Observatory*, **87**, 295.
Lamers, H.J.G.L.M., de Groot, M., & Cassatella, A. (1983a). *Astr. Ap.*, **123**, L8.
Lamers, H.J.G.L.M., de Groot, M., & Cassatella, A. (1983b). *Astr. Ap.*, **128**, 299.
Lamers, H.J.G.L.M., Korevaar, P., & Cassatella, A. (1984). *Proc. 4th European IUE Conference, ESA SP-218*, p. 315.
Lamers, H.J.G.L.M., Korevaar, P., & Cassatella, A. (1985). *Astr. Ap.*, in press.
Lamers, H.J.G.L.M. (1985). In *Luminous Stars in Associations and Galaxies*, ed. C. de Loore & A. Willis, in press.
Luud, L.S., Gollandsky, O., & Yarygina, T. (1975). *Pub. Tartu Astrofiz. Obs.*, **43**, 250.
Maeder, A. (1981). *Astr. Ap.*, **99**, 97.
Maeder, A. (1983). *Astr. Ap.*, **120**, 113.
Magalashvili, N.L. & Kharadze, E.K. (1967a). *Astron. Tsirk*, No. 426.
Magalashvili, N.L. & Kharadze, E.K. (1967b). *Comm. 27 IAU Inf. Bull. Var. Stars*, No. 210.
Markova, N. & Kolka, I. (1984). *Astrofiz*, **20**, 465 (= *Astrophys.*, **20**, 250).
Morgan, W.W., Johnson, H.L., & Roman, N.G. (1954). *Pub. A. S. P.*, **66**, 85.
Nikonov, B. (1937). *Bjull. Abastumani Astrofiz. Obs.*, **1**, 35.
Nikonov, B. (1938). *Bjull. Abastumani Astrofiz. Obs.*, **2**, 23.
Percy, J.R. & Welch, D.L. (1983). *Pub. A. S. P.*, **95**, 491.
Pigott, E. (1786). *Phil. Trans. R. Soc. London*, **76**, 189.
Schneller, H. (1957). *Geschichte und Literatur des Lichtwechsels der Veränderlichen Sternen*. Berlin: Akademie-Verlag.
Smeets, H.J.H., Heintze, J.R.W., & van Gent, R.H. (1985). in preparation.
Stahl, O., Wolf, B., Zickgraf, F.-J., Bastian, U., de Groot, M.J.H., & Leitherer, C. (1983). *Astr. Ap.*, **120**, 287.
Stahl, O., Wolf, B., Leitherer, C., Zickgraf, F.-J., Krautter, J., & de Groot, M.J.H. (1984). *Astr. Ap.*, **140**, 459.
Van Genderen, A.M. (1979a). *Astr. Ap. Suppl.*, **38**, 151.

Van Genderen, A.M. (1979b). *Astr. Ap. Suppl.*, **38**, 381.
Van Gent, R.H. & Lamers, H.J.G.L.M. (1985). *Astr. Ap.*, in press.
Wendker, H.J. (1982). *Astr. Ap.*, **116**, L5.
Wolf, B., Stahl, O., de Groot, M.J.H., & Sterken, C. (1981a). *Astr. Ap.*, **99**, 351.
Wolf, B., Appenzeller, I., & Stahl, O. (1981b). *Astr. Ap.*, **103**, 94.

THE RAPIDLY OSCILLATING Ap STARS

Jaymie M. Matthews
Department of Astronomy
University of Western Ontario
London, Ontario
Canada N6A 3K7

I. INTRODUCTION

Light variations with periods ranging from four to 15 minutes and amplitudes typically less than $0.^m01$ have been convincingly detected in eleven cool Ap stars to date (*e.g.*, Kurtz 1982). Since these stars are relatively bright, and there are many comparably bright candidates yet to be searched for this type of variability, the study of the ***rapidly oscillating Ap stars*** ("roAp stars") is well suited to small telescopes. In fact, the number of weeks required to adequately monitor each variable leaves little choice but to use such telescopes, given the hard reality of observing time allocations on large-aperture instruments.

The roAp stars exhibit many interesting features. For example, the amplitude of oscillation is modulated in phase with the magnetic (rotational) period of the star (when measured), such that the maximum amplitude coincides with greatest field strength. (The peculiar A, or Ap, stars have the most intense magnetic fields yet detected in normal stars, sometimes as high as tens of kilogauss (Borra & Landstreet 1980).) Also, when the light curve is analyzed using Fourier techniques (see article by Fullerton elsewhere in this volume), the peaks which represent the observed frequencies are split into equally spaced components. The frequency spacing corresponds to the period of the magnetic variation, usually several days for the Ap stars.

II. THE "OBLIQUE PULSATOR" AND OTHER MODELS

The empirical properties of the roAp stars can be explained, at least in part, by the ***oblique pulsator*** hypothesis proposed by Kurtz (1982). This model is itself based upon another model, the ***oblique rotator*** (Stibbs 1950), which has been very successful in accounting for the *long-term* magnetic, spectroscopic, and photometric variations seen in Ap stars. According to the oblique rotator model, an Ap star possesses a dipole magnetic field (not unlike the configuration of the Earth's field) whose axis is tipped with respect to the axis of rotation. As the star rotates, the magnetic poles move relative to the observer's line of sight, resulting in periodic variations of the apparent field strength. At the same time, "patches" of different atmospheric abundances and brightness — conforming to the surface geometry of the magnetic field — are also carried across the stellar disc by rotation. Therefore, the magnetic, spectroscopic, photometric, *and* rotation periods of an oblique rotator are identical, as is observed in many Ap stars.

In the oblique pulsator model, a roAp star is an oblique rotator which is also pulsating nonradially, but with the pulsation pattern symmetric about the magnetic axis. Just as the observed magnetic field strength is modulated with rotation, so is the pulsation amplitude, and with the same period. The model also predicts the frequency splitting which is resolved in the Fourier spectra of these stars' oscillations.

A simple case of the oblique pulsator is illustrated in Figure 1. The star depicted is assumed to be pusating in an $l = 1$, $m = 0$ mode, while rotating at a moderate inclination. (The shaded portion represents a hemisphere whose surface motions are 180° out of phase with the other half of the star. The boundary is a *node* where no motion occurs.) The north and south poles of a dipole magnetic field, tilted 45° to the axis of rotation, are labelled; these are also taken to be the pulsation "poles". The schematic graph shows the modulation of the observed oscillation amplitude during the star's rotation, reaching maximum amplitude when a magnetic pole is most directly pointed towards the observer. Note that the phase of the oscillation also jumps sharply as the different pulsation hemispheres dominate the visible disc. Both of these effects have been recorded in the roAp stars.

A competing model, the *spotted pulsator* (Mathys 1984), suggests that the amplitude modulation instead arises because of the surface brightness anomalies associated with Ap stars. Unfortunately, this model has so many free parameters that it is difficult to test at this time.

Although both of these models implicitly assume nonradial pulsation as the source of the variations, the actual mechanism is still unknown. Since the rapid oscillators fall in or near the lower instability strip on the HR diagram, it is attractive to consider them as pulsators driven by the same envelope ionization mechanisms which govern the δ Scuti variables. The short periods of the roAp stars are more difficult to explain in this context; the strong magnetic fields of the Ap stars have been invoked to somehow filter out the lower "δ Scuti-type" overtones of pulsation. However, there is no theoretical basis for this as yet. Another proposed mechanism, *magnetic overstable convection* (Shibahashi 1983), also calls upon the magnetic field to help explain the oscillations. In this instance, the field lines act like elastic bands "glued" to the plasma in the stellar atmosphere. They provide a restoring tension against convective motions, which can induce vibrations with the correct timescales.

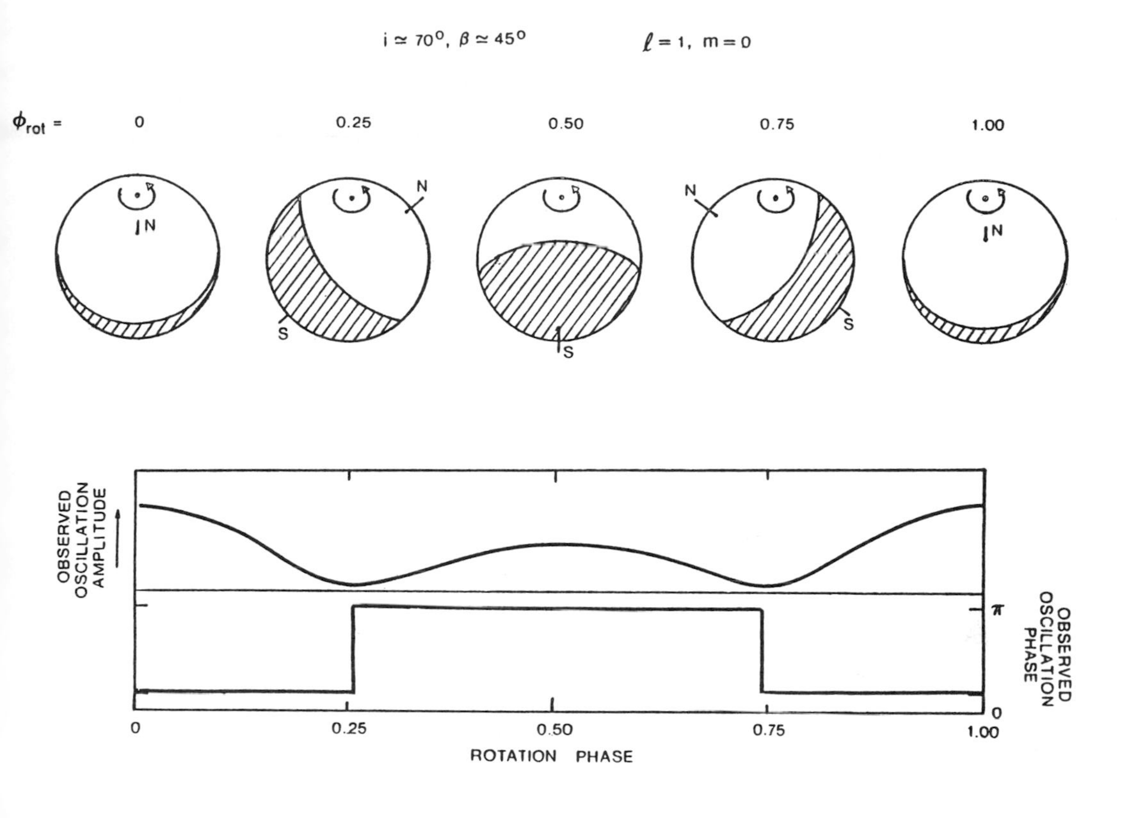

Figure 1. The oblique pulsator model for rapidly oscillating Ap stars. (See text for explanation.)

III. OBSERVATIONS OF THE roAp STARS AND THE ROLE OF SMALL TELESCOPES

Some parameters for ten of the known roAp stars are presented in Table 1. (The eleventh member of the class, HD 134214, was only recently discovered at the time of writing. Tobias Kreidl (private communication), using the 1.1 m telescope of the Lowell Observatory, found that this star oscillates with a period of about 5.65 minutes and a peak B amplitude near 6 mmag.)

The roAp stars are challenging objects for the photometric observer. Since observations are acquired at the rate of one every few seconds, it is almost essential to record them automatically on diskette or magnetic tape. The detection of oscillations with periods as short as four minutes or less, using a conventional single-channel photometer, requires *non*-differential photometry (*i.e.*, the variable is

Table 1. The Rapidly Oscillating Ap Stars

HD HR	Name	(b-y)	T_{eff} (K)	Periods (min)	ΔB(max) (mmag)	H_e (gauss)	sp type
6532		0.084	8000-8800	6.94 (14.28?)	2.5		A5pSrCrEu
24712 1217		0.183		6.14	14.5	+300 to +1200	F0pSrCrEu
60435		0.132		11.9 15.2 6 4.0	15.7		F0pSrEu
83368 3831		0.146		11.67 5.835	10.2	-700 to +700	F0pSrEuCr
101065	Przybylski's star	0.448	7400	12.14	14.6	-2200	
128898 5463	α Cir	0.152		6.825	4.8	-300	F0p
137949	33 Lib	0.188	8000-8800	8.27	2.8	+1400 to +1800	F0pSrCrEu
201601 8097	γ Equ	0.147	8050	12.5	2.8	+500 to -800	F0p
203932		0.169		5.94	2.4		A5p
217522		0.289		13.72	4.2		A5p

observed continuously, without reference to a comparison star). Because of this and the very small amplitudes characteristic of these variables, stable sky transparency is a must for useful measurements.

Still, changes in transparency — neglecting the systematic rise and fall with zenith angle due to air mass extinction (which can usually be removed from the data during reduction) — are inevitable, even on nights of excellent quality. Luckily, those fluctuations tend to be gradual, of low amplitude, and random in nature. The rapid coherent oscillations of a roAp star can be distinguished from most sky variations by Fourier analysis of the light curve.

Even without allowance for variable extinction, a study of stellar variations in the frequency domain often yields accuracy and detail unattainable in the time domain. Figure 2 demonstrates the value of Fourier analysis to the task of frequency identification. Figure 2*a* is a typical light curve of the roAp star HD 60435, studied extensively by Matthews *et al.* (1985). Neither the oscillation amplitude nor the observing conditions were extraordinary on this night. Although there are strong indications in the data of a rapid oscillation, particularly near the end of the run, the situation is not very clear. The corresponding *periodogram* (a discretely sampled estimate of the Fourier amplitude spectrum) in Figure 2*b* unambiguously reveals the dominant oscillation, with a period near 12 minutes. There is a wealth of information which can be extracted from a light curve by such methods.

The roAp stars also present serious logistic problems to the prospective observer. Since the amplitudes of oscillation are modulated over timescales of days, many nights are necessary to determine the modulation period and resolve the attendant fine structure in the Fourier spectrum. In addition, cycle/day aliases or false peaks in that spectrum — arising from the daily gaps inherent in optical astronomical observations — confuse what can be an already complicated pattern of frequencies (see Figure 3). The only way to deal effectively with all of these obstacles is to observe the star from two or more observatories, spaced widely in longitude, over an interval of at least several weeks. The advantages are two-fold: i) The chance of at least one observatory having useable skies during each night increases with the number of sites. ii) The gaps between segments of continuous data are reduced, along with the sizes of the corresponding alias peaks in the frequency spectra.

In light of the heavy demand in the astronomical community for large telescope time, an extended multi-site observing campaign is only practical using small or moderately-sized telescopes (see article by Chris Sterken elsewhere in this volume). Fortunately, there are many bright ($V \lesssim 9$) Ap stars for which good photon statistics can be obtained using, say, a 0.5 m telescope and integration times of only 10 - 20 seconds.

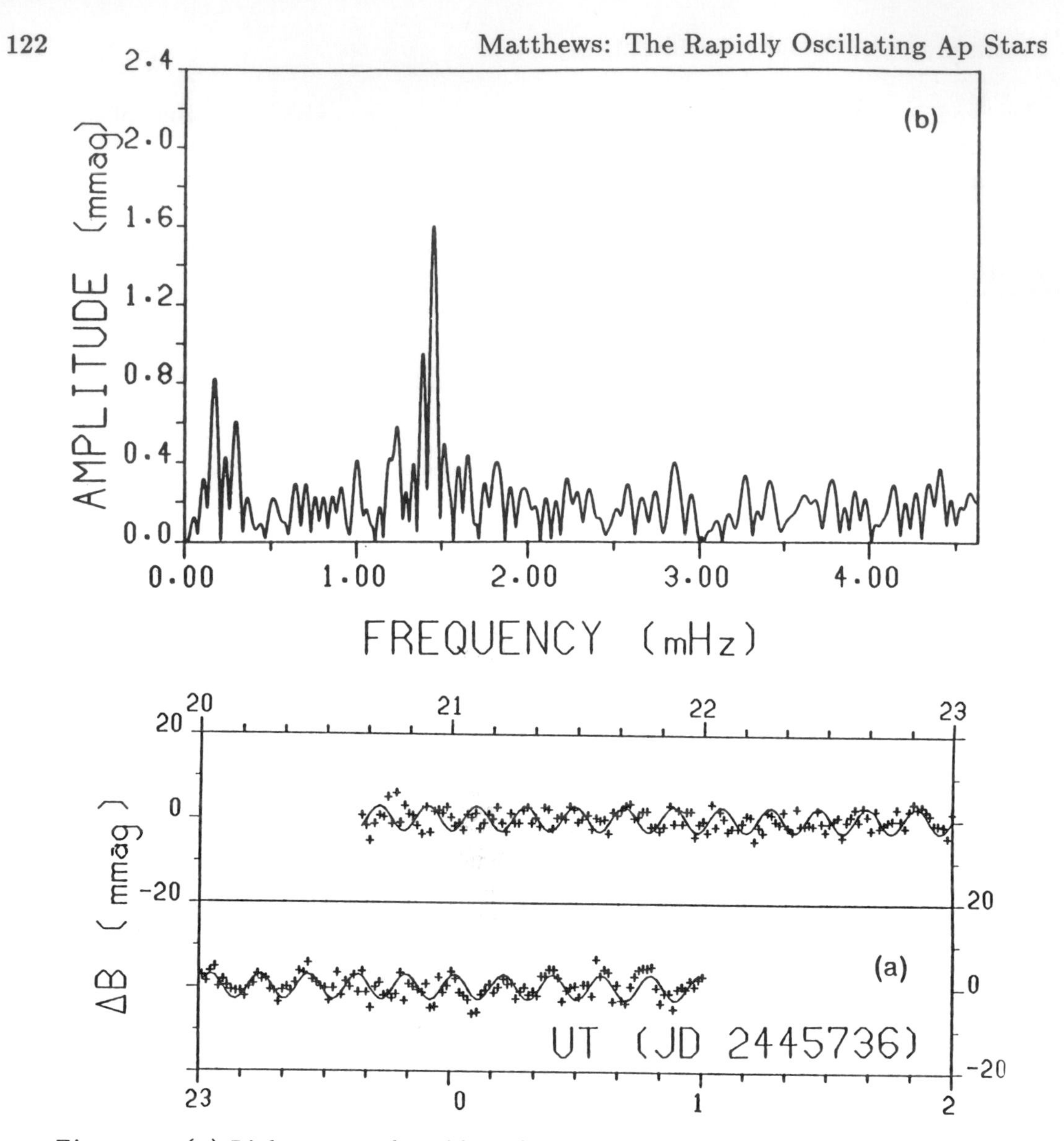

Figure 2. *(a) Light curve of rapid B photometry of HD 60435, obtained by D.W. Kurtz using the 0.5 m telescope of the South African Astronomical Observatory (SAAO). The crosses represent three-point averages of 20 second integrations. No comparison star was employed. A sinusoid with a period of 11.8 minutes has been superimposed on the data. (b) Periodogram of the light curve in (a). The largest peak occurs at a frequency near 1.4 mHz (period = 11.8 minutes). A second smaller peak is resolved just shortward in frequency. This hints at the more complicated frequency pattern revealed by later detailed analysis. (Figures 2b and 3 are taken from Matthews et al. (1985).)*

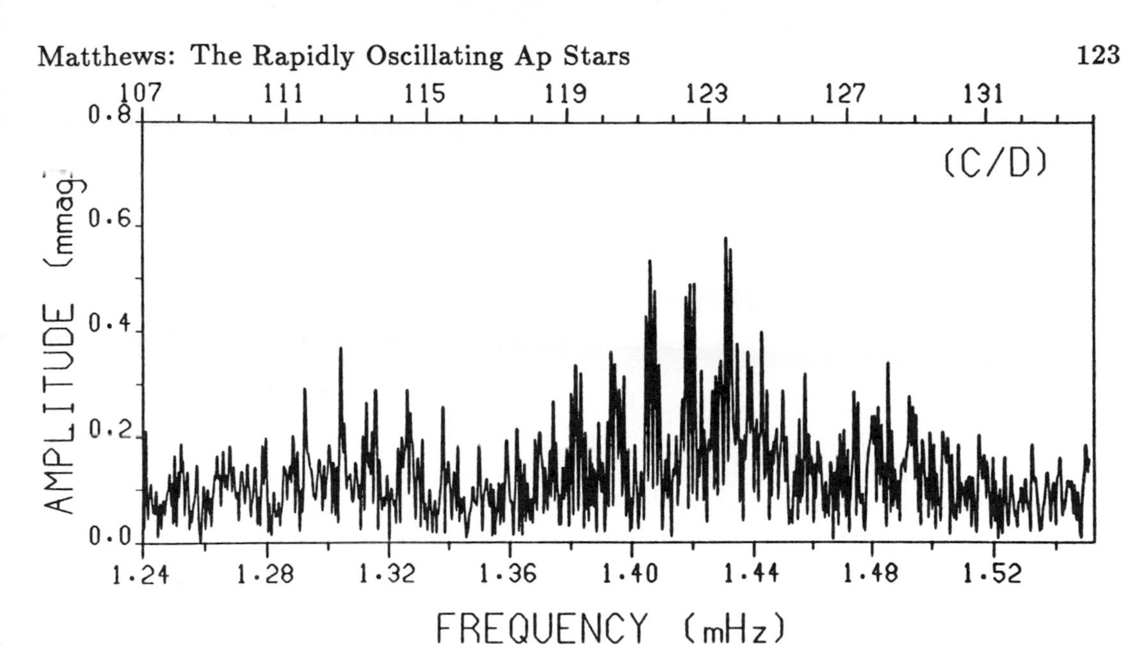

Figure 3. Periodogram of 18 nights of data of HD 60435, collected from the University of Toronto 0.6 m telescope on Las Campanas, Chile, and the SAAO 0.5 m telescope. Only the frequency region near 1.4 mHz is shown. Even with some overlapping observations from the two sites, the cycle/day aliases and the inherent complexity of the oscillations make frequency identification difficult. There are probably only four "true" peaks above the noise level in this spectrum. (This is based on comparison to spectra of data without daily aliases, and filtering of the known aliasing pattern from the periodogram.) The other peaks are presumed to be aliases or rotationally-split components.

IV. CONCLUSION

Comprehensive study of even a single rapidly oscillating Ap star demands many weeks of photometry — with high time resolution and accuracy — from several sites, followed by intensive frequency analysis. The effort is consuming of both man- and computer-hours, but the potential returns are certainly worth the investment.

The significance of the roAp stars extends far beyond simply the study of another new class of variable. The discovery of the "5-minute" and related low-amplitude pulsations in the sun (Leighton *et al.* 1962) has since spawned a new field: *helioseismology.* Like geophysicists, who have been able to model the Earth's interior through careful monitoring of its seismic vibrations, astronomers can now learn about the solar interior by comparing the frequencies and spacings of the sun's surface oscillations to the predictions of theoretical models. The same technique, applied to the frequency patterns of the rapid oscillators (if they are indeed pulsating stars) shows great promise in improving our understanding of the interior structure of the Ap stars.

Already, the frequency spectra of roAp stars such as HR 1217 (Kurtz & Seeman 1983) and HD 60435 (Matthews *et al.* 1985) have been compared to the models of main sequence A stars developed by Shibahashi & Saio (1985) and Gabriel *et al.* (1984). The preliminary results of such *astroseismology* suggest that the roAp stars must have abnormal temperature *vs.* optical depth profiles to permit pulsations with the observed high frequencies. Continued observations and refined models of these stars are called for.

It seems surprising that optical photometry obtained with ground-based telescopes as small as 0.5 m in aperture may provide key insights into the interiors of distant stars!

REFERENCES

Borra, E.F. & Landstreet, J.D. (1980). *Ap. J. Suppl.*, **42**, 421.
Gabriel, M., Noels, A., Scuflaire, R., & Mathys, G. (1984). *Astr. Ap.*, **143**, 206.
Kurtz, D.W. (1982). *M. N. R. A. S.*, **200**, 503.
Kurtz, D.W. & Seeman, J. (1983). *M. N. R. A. S.*, **205**, 11.
Leighton, R.B., Noyes, R.W., & Simon, G.W. (1962). *Ap. J.*, **135**, 474.
Matthews, J.M., Kurtz, D.W., & Wehlau, W.H. (1985). *Ap. J.*, in press.
Mathys, G. (1984). In *Theoretical Problems in Stellar Stability and Oscillations*, 25th Liege International Astrophysical Colloquium.
Shibahashi, H. (1983). *Ap. J. (Letters)*, **275**, L5.
Shibahashi, H. & Saio, H. (1985). *Pub. Astr. Soc. Japan*, in press.
Stibbs, D.W.N. (1950). *M. N. R. A. S.*, **110**, 395.

IV. OTHER TECHNIQUES

The papers in this section deal with techniques other than the standard, broad-band measurement of visible light — techniques not usually associated with most small observatories. Until recently, infrared photometry required special expertise and equipment, and spectroscopy (due to the limited sensitivity of the photographic plate) required a moderate to large telescope. Now, near-infrared photometers can be bought inexpensively, and sensitive solid-state spectroscopic detectors are within the budget of many small observatories. Nevertheless, although these techniques are accessable to many small observatories, they are not done casually — they are best done by an individual or an institution willing to pursue them seriously over a significant period of time.

Robert F. Wing has devoted much of his career to infrared studies of Mira variables and other cool stars. Through his collaboration with Gerald Persha, he has helped to bring infrared techniques within reach of all photometrists. His review provides a clear account of the science of near-infrared photometry, and introduces the notion that, by observing at two or more wavelengths, one can measure more than just brightness. This is the essence of spectroscopy. Douglas L. Welch, whose interest in photometry began when he was an amateur astronomer in high school, and has continued through his years as an undergraduate and graduate student, demonstrates in his paper how a large and exciting project in extragalactic near-infrared astronomy depends on small-telescope astronomy for its roots.

Robert F. Garrison, a world leader in the field of stellar spectral classification, serves as director of the University of Toronto Southern Observatory at Las Campanas in Chile, arguably the most productive small telescope in the world. He describes the site and some of the research programs of the observatory, emphasizing the advantages of a telescope whose scheduling policies are more flexible than those of the national observatories. This is followed by a paper by Stanley Jeffers and William G. Weller, which gives an example of how such a telescope, when equipped with a sensitive detector, can be used to study spectroscopic variability.

OBSERVATION OF VARIABLE STARS IN THE INFRARED

Robert F. Wing
Astronomy Department, Ohio State University
Columbus OH 43210
USA

ABSTRACT

Infrared photometry of cool variable stars is a field that holds considerable promise for observers with small telescopes. Following general remarks about the infrared, recommendations are given for long-term observing programs involving (a) the measurement of infrared light curves of Mira variables and (b) the photometric measurement of spectral types.

An interesting recent development in observational astronomy — but one which is not yet widely recognized — is that the infrared part of the spectrum is becoming more accessible to observers with small telescopes and limited budgets. Infrared photometry of variable stars, which to date has been done almost exclusively by specialists working at the major observatories, is thus moving into the province of the small observatory and even the amateur. Since many of the stars bright enough to be studied effectively with small telescopes are red giants — stars which are particularly bright in the infrared and which are often variable — the participation of amateurs and student groups in this field is likely to have an important impact on variable-star research.

My purpose today is to encourage infrared photometry of red variables not only by astronomers already equipped for such work but also by other observers who may be attracted to an exciting but straightforward line of research. I will do this by describing some fairly specific observational programs which would benefit from the participation of a large number of observers. But first I should dispell a few myths concerning infrared observing in general.

I. GENERAL REMARKS ABOUT THE INFRARED

It is still widely believed that observational work at infrared wavelengths can be done only by astronomers of a rare breed — known as "infrared astronomers" — who have highly specialized training and access to special equipment mounted on large telescopes at dry, high-altitude sites. Consequently, few individuals from outside this group have considered doing infrared work themselves. But those who have investigated the matter have found that the difficulties implied above apply mainly to work at wavelengths beyond 3 or 4 microns (μm), in the region sometimes loosely called the "far infrared". But in the broad and important interval stretching from the long-wavelength end of the visible region at about 0.75 μm out to at least 2 μm, these concerns are much less important.

Consider, for example, the altitude of the observatory. If high altitude were a strict requirement it would eliminate nearly all amateur observers and college observatories from participation, and the other factors would be academic. Indeed, for work beyond 3 or 4 μm it is important to get above as much of the atmospheric water vapour as possible, since this introduces "sky noise" as well as absorption which is likely to be time-variable. High-altitude sites are often called "infrared sites", and site surveys for next-generation telescopes are concentrating on sites above 3,000 m for precisely this reason. But for measurements at wavelengths shorter than about 3 μm, the altitude of the observatory is hardly more important a consideration than it is for photometry in the visible region. To be sure, there are absorption bands of water vapour scattered through the near infrared, but they can be avoided quite effectively with appropriate filters since the windows between the bands are reasonable clear. I see no reason why good photometry at 2 μm should not be possible from sea level, provided that certain precautions are taken.

In Table 1 the characteristics of three infrared wavelength regions are compared. The first, from 0.75 to 1.10 μm, is accessible to photoemissive devices and solid-state detectors such as silicon diodes. The second, from 1.2 to 2.4 μm, is the region of the broadband J, H, and K filters and is usually studied with photovoltaic (PbS or InSb) detectors. The third region, from 3 to 20 μm, is also accessible from the ground but only with great difficulty. To these could be added a fourth region, observable only from space, defined as the range surveyed recently by the Infrared Astronomical Satellite (IRAS), which extends to 100 μm. The three regions of Table 1 are often called the "near", "intermediate", and "far" infrared, but I will avoid using these terms because their meaning depends upon the context. For example, astronomers using IRAS data are apt to think of the "near infrared" as extending as far as 20 μm.

An important change occurs at a wavelength of about 3 or 4 μm: below this wavelength the sky is dominated by ordinary stars, but at longer wavelengths the brightest objects are cool extended regions such as nebulae and circumstellar shells and solid bodies such as asteroids. Even if one specifically observes ordinary stars in the long-wavelength region, analysis of the energy distribution usually shows that most of the radiation detected comes not from the star's photosphere but from dust grains in a large shell of circumstellar material. Thus reseachers working longward or shortward of 3 μm are likely to be interested in completely different classes of problems.

Table 1. Comparison of infrared wavelength regions.

	0.75 – 1.10 μm	1.2 – 2.4 μm	3 – 20 μm
Principal objects ...	............Stars	....	Shells, nebulae
Brightest star	..........Betelgeuse	....	IRC+10216
Principal detector .	Photomultiplier; Si photodiode	Photovoltaic cell	Bolometer
Special requirements	........Filters to avoid atm. H_2O bands	....	High-altitude site chopping secondary liquid He cooling

The well-known star Betelgeuse (α Ori) is the brightest star in the sky from about 0.75 μm (where it overtakes Sirius) to about 3 μm, *i.e.*, throughout our first two regions. At 2 μm its closest competitor is R Dor, an M8 giant in the southern sky (see Wing (1971) for lists of the brightest stars at 1 and 2 μm). At still longer wavelengths, the distinction of being the brightest object outside the solar system passes to IRC+10216 (sometimes called "Becklin's star"), a carbon star that is completely enshrouded in a thick envelope of circumstellar dust. Nearly all the radiation that we receive from it is featureless thermal emission from the dust grains, rather than radiation from the star's photosphere.

Photometry longward of 3 μm places extraordinary demands on the observer and his equipment. The detector must be cooled to an extremely low temperature, usually with liquid helium. Thermal radiation from the telescope optics and the atmosphere would swamp the signals from celestial sources without an elaborate mechanism for "chopping" between the source and the adjacent background. Work at these long wavelengths is thus best done with "IR-optimized" telescopes with oscillating secondary mirrors.

In the 1 μm and 2 μm regions, on the other hand, any telescope designed for visual observations can be used. Photoemissive detectors used in the 1 μm region need to be cooled with dry ice, but silicon diodes and photovoltaic cells can be used effectively at the ambient temperature. Some arrangement for rapid sky subtraction is highly desirable, but a focal-plane chopping mechanism is adequate here and much easier to provide than the oscillating secondary mirror needed for work at longer wavelengths.

The obstacles to working beyond 3 μm would appear to exclude the participation of small observatories. But this should in no way discourage potential new infrared observers, since photometry can be obtained as far into the infrared as 2 μm with equipment and techniques that are not greatly different from those used for photoelectric photometry in the visible region. Interested observers will find a number of useful articles in recent issues of the *IAPPP Communications*, especially *Communication* No. 14 (Wolpert, Hall, & Reisenweber 1983) which is a special issue on infrared and solid-state photometry. A K-band photometer with a PbS detector is being manufactured by Optec, Inc. and is well suited for the measurement of infrared light curves of Mira variables (see next section). Also Persha (1986) has shown that the Optec SSP-3 photometer, which employs a silicon photodiode and is normally used for $BVRI$ photometry, can also be fitted with special filters to measure narrow-band indices of TiO strength, the uses of which are discussed in the final section of this paper.

II. INFRARED LIGHT CURVES OF MIRA VARIABLES

The Mira variables are, of course, well known to all variable-star observers and particularly to members of the AAVSO. Their variations are sufficiently regular that observing programs can be planned in advance, yet not so regular as to make continued monitoring unnecessary. Their large amplitudes make the Miras attractive to visual observers since relatively large errors in magnitude can be tolerated in determining their light curves. Their periods of several hundred days are rather long for coverage by an individual observer but quite convenient for an organization coordinating the efforts of many observers. Photoelectric light curves have been published for a few Miras in a few cycles, but most of what we know of the light curves of Miras has come from the visual observations collected by the AAVSO and sister organizations overseas.

In the infrared, the first thing one notices about Miras is how bright they are. They are among the coolest of all stars, and their energy maxima generally lie between 1 and 2 μm in the infrared. Typical Miras are anywhere from 6 to 10 magnitudes brighter in this region than they are visually, making them ideal targets for infrared photometrists with small telescopes. There is something very satisfying about recording a strong signal from a star that is barely visible in the eyepiece, a common experience for infrared observers of Mira variables. For one thing, you don't need to wonder whether you're observing the right star!

The infrared magnitude of a Mira variable varies with the same period as the visual magnitude, but with a much smaller amplitude and noticeably different phasing. The characteristics of the infrared light curves of Miras were first determined by Pettit & Nicholson (1933), who used a bolometer at the Mount Wilson 100 inch to record all the radiation passing through the Earth's atmosphere. They found that the bolometric magnitude (*i.e.*, the magnitude representing the total flux, summed over all wavelengths) reaches maximum brightness long after the visual maximum, the phase lag amounting to 10 - 15 per cent of the period (or typically 1 - 2 months).

The bolometric light curve directly reflects the star's changes in effective temperature and radius (or surface area), and it is more useful to theoreticians studying the star's pulsations than a host of light curves at discrete wavelengths. We must therefore consider how the bolometric magnitude can be determined. One can, like Pettit and Nicholson, use an unfiltered bolometer which responds equally to radiation of all wavelengths, but then there is no simple way to correct for the absorption by the Earth's atmosphere (for which one needs to know the star's detailed energy distribution). Modern determinations of m_{bol} (e.g., Smak 1966) use multicolour wideband infrared photometry and interpolations based on balloon or satellite scans of representative stars to allow for the parts of the spectrum missed by the filters. Both procedures are difficult, but fortunately it turns out that the bolometric light curves derived by either technique do not differ significantly from single-filter curves obtained for any continuum filter (*i.e.*, one that avoids strong stellar molecular absorption bands) in the 1 - 2 μm region around the flux maximum. In particular, the narrow-band $I(104)$ magnitude at 1.04 μm and the wideband K magnitude at 2.2 μm both give good representations of the bolometric magnitude.

Bolometric and visual light curves for a typical Mira variable are shown schematically in Figure 1. For stars with normal (M-type) spectra the amplitude of the bolometric curve is about five times smaller than the visual amplitude; here the amplitudes have been drawn as 1.0 mag and 5.0 mag, respectively, as these are the mean $I(104)$ and V amplitudes of the 25 stars studied by Lockwood & Wing (1971). This means that a Mira is typically only 2.5 times fainter at minimum than at maximum in the infrared, whereas visual observers have to contend with a factor of 100.

In the V curve, the rise to maximum is more rapid than the decline, so that minimum light occurs approximately 0.6 cycle after visual maximum. The infrared (or bolometric) curve is more symmetrical but is displaced by a phase lag of 0.10 or 0.15 cycle relative to the visual light curve.

To understand why the visual light curve of a typical Mira differs so much from the bolometric light curve, we must consider how the visual region differs from a representative region of the infrared continuum. If Miras were simply blackbodies of variable temperature, the amplitude at a relatively short wavelength such as V would be larger than the bolometric amplitude, but not by nearly as large a factor as is observed; furthermore, the curves for different filters would all be in phase. Emission lines, which occur throughout the spectra of Miras and which come and go

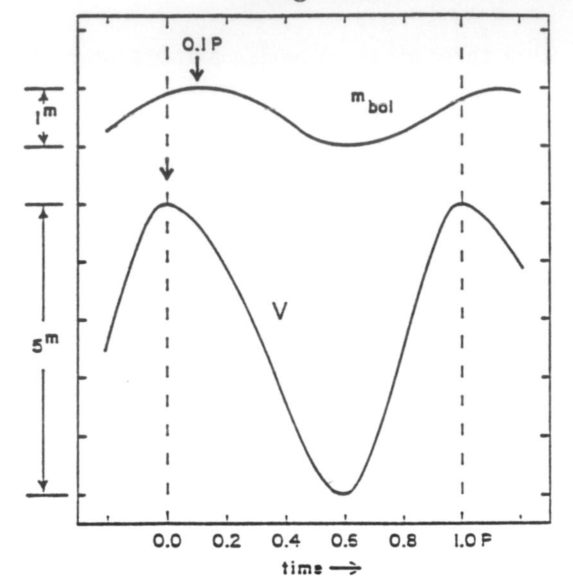

Figure 1. Schematic visual and bolometric light curves for a typical Mira variable. The bolometric curve has one-fifth the amplitude of the visual curve and a phase lag of about one-tenth of the period. Light curves measured in the infrared continuum have approximately the same amplitude and phasing as the bolometric curve.

with phase, are relatively unimportant in the visual region. What *does* matter are the variations in the strength of the absorption bands of the TiO molecule, which cover the entire visual region and are very sensitive to changes in temperature. Several years ago, J. Smak and I used calibrated spectral scans from the ultraviolet to the near infrared to determine the amount of absorption caused by TiO in each of the filters B, V, R, and I as a function of the spectral type (*i.e.*, TiO strength) of the star (Smak & Wing 1979). Table 2 gives our results in abbreviated form. Now, a typical M-type Mira might vary from type M4e at maximum to type M8e at minimum, and we see from the table that the blanketing in the V filter changes by 1.80 mag in that case. The star's visual amplitude is thus 1.8 mag greater than it would be from the temperature change alone. The table shows that the amplitudes in B, R, and I are also enhanced by the changes in TiO absorption, although by lesser amounts.

Analyses of the spectra and colours of Miras have confirmed the conclusion of Pettit and Nicholson that visual maximum corresponds to the time of highest temperature and weakest TiO band strength (*i.e.*, earliest spectral type). The temperature sensitivity of the TiO bands acts to increase the visual amplitude and to ensure that visual maximum occurs at the time of maximum temperature. To account for the fact that the bolometric magnitude is *not* brightest at the time of

Table 2.* Effect of blanketing by TiO.
(in magnitudes)

Spectral Type	B	V	R	I
M0	0.00	0.00	0.00	0.00
M2	0.25	0.30	0.20	0.05
M4	0.65	0.95	0.55	0.15
M6	1.20	1.85	1.05	0.30
M8	1.75	2.75	1.55	0.50

* from Smak & Wing (1979)

highest temperature, we must conclude that the star is smallest around that time, and that it swells greatly during the first few months after visual maximum. A non-technical review of the pulsational properties of Miras, which reconciles the apparently conflicting evidence from spectroscopic and photometric data, was prepared by the writer a few years ago (Wing 1980).

Individual Mira variables, of course, differ considerably in the shapes of their mean visual light curves. The shorter-period Miras (P $\sim$ 200 days) have quite uniform shapes, but the longer-period ones (P $\sim$ 400 days) are more varied. In Figure 2 we show the mean light curves determined by the AAVSO for six long-period Miras, all with periods close to 410 days (Campbell 1955). Since these stars have a wide variety of spectral characteristics (including types M, S, and C) as well as an assortment of light-curve forms, we can ask whether the differences in their light curves can reasonably be attributed to spectroscopic effects, *i.e.*, to changes in the blanketing of the visual region by TiO bands.

One positive result can be stated with confidence: carbon stars, such as WX Cyg in Figure 2, have relatively small visual amplitudes because their spectra lack the temperature-sensitive metallic oxide bands which amplify the variations of M and S stars. The chief absorption features in the visual spectra of carbon stars are bands of C_2 and CN which are not particularly sensitive to temperature, so that they have little effect on the shape of the light curve. Support for this interpretation can be found by inspection of the other mean light curves for Mira stars given by Campbell (1955): it is actually quite easy to pick out the carbon stars on the basis of their visual light curves alone.

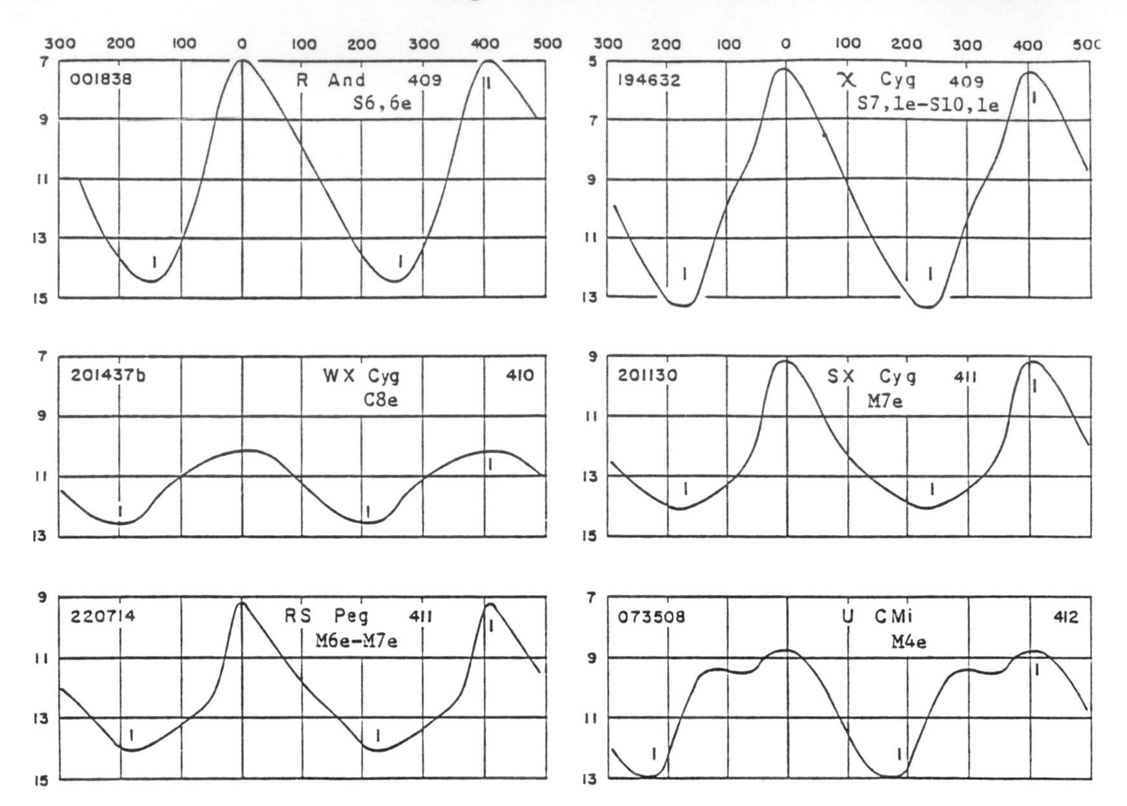

Figure 2. Mean visual light curves for six Mira variables, from Campbell (1955). The spectral types have been added under the star names; note that representatives of types M, S, and C are included.

On the other hand, other differences in the light curves seem harder to associate with differential blanketing effects and may have nothing to do with them. The occasional occurrence of a "hump" on the rising branch is a good example. The star U CMi, which shows a prominent hump, has a normal M-type spectrum, and many stars which seem spectroscopically indistinguishable from U CMi show no sign of a hump at all.

The best way to determine whether the peculiarities of visual light curves are caused by spectroscopic effects in the visual filter is to measure light curves in the infrared, using filters that isolate relatively clean continuum regions. If humps or other features of the visual light curve also appear in the infrared continuum (or bolometric) light curve, they are obviously not caused by spectral changes in the visual region and must be explained in terms of the phasing of the temperature and diameter variations.

A first attempt to investigate this question was made when G. W. Lockwood and I pooled our measurements of $I(104)$ magnitudes of Miras (Lockwood & Wing 1971). We found that we had fairly extensive data for 25 stars covering several cycles. We recorded several instances of humps on the rising branches of $I(104)$ light curves (see Figure 4 below) and concluded that this phenomenon is *not* caused by differential blanketing effects.

Another well-known property of the visual light curves of Miras is their failure to reproduce themselves from cycle to cycle. It is not uncommon for the visual magnitude at the maximum of a particular cycle to differ from the star's mean value of V_{max} by a full magnitude or more, and individual cycle lengths are typically several per cent longer or shorter than the mean period. One can get a good idea of these cycle-to-cycle differences from *AAVSO Report 38* (1983), which presents plots of the individual magnitude estimates collected during a three-year interval for several hundred variables. Do these cycle-to-cycle differences also occur in the infrared continuum? If so, they must also occur in the bolometric light curve and must reflect a degree of randomness in the temperature curve and/or the diameter curve. But if the infrared light curves are consistent from cycle to cycle, the differences in the V curves might better be explained as the result of more superficial effects such as differences in TiO blanketing.

During my graduate student days at Lick Observatory, I observed the 1965 and 1966 cycles of W Peg, a normal M-type Mira, with UBV photometry and near-infrared scanner measurements. In Figure 3 the $I(104)$ and V light curves are shown. The two cycles are indistinguishable in the infrared continuum despite significant differences in the V curves. The spectral type obtained from scanner measurements of TiO bands showed that the differences in the V curves could be entirely accounted for by differential blanketing effects. If I had observed only this one star in these two cycles, I might have announced the satisfying — but totally incorrect — result that the infrared light curves of Miras are the same in every cycle and hence need only be observed once. Fortunately I had other data which warned me not to make sweeping generalizations, and by the time that Lockwood and I put together six years' worth of data in 1971 it was clear that substantial cycle-to-cycle differences in the $I(104)$ curves are the rule rather than the exception.

Some examples of light curves in $I(104)$ are reproduced from Lockwood & Wing (1971) in Figure 4. Observations from different cycles have been combined using the mean period but are kept distinct by means of different symbols and connecting lines. Humps on the rising branch are quite common in this sample of stars, which is believed to be representative. Cycle-to-cycle differences are substantial, exceeding 0.5 mag in the case of χ Cyg and R And. We see that continued monitoring is just as necessary in the infrared as it is in the visual.

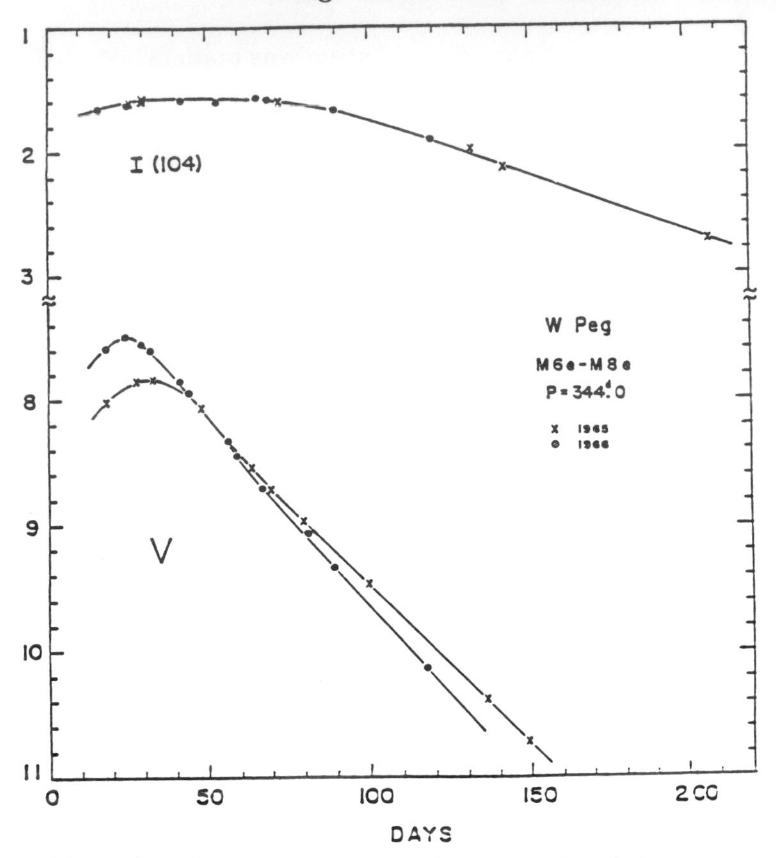

Figure 3. Photoelectric measurements of V and I(104) for W Peg in two successive cycles. The infrared continuum light curves are indistinguishable. From Wing (1967).

Observers equipped to measure $I(104)$ or K magnitudes can make an important contribution to the study of Mira variables by monitoring these infrared continuum magnitudes, thereby providing a good approximation to the bolometric light curves. This has never been done for a large number of stars, or for an extended period of time. Surely the bolometric light curves deserve as much attention as the visual, and a beginning can now be made with the relatively simple and inexpensive equipment for infrared photometry that has recently become available. Meanwhile, of course, the work of visual observers must continue, not only for continuity but also because much can be learned from the comparison of visual and infrared light curves.

The amount of work waiting to be done along these lines is virtually unlimited, and a great deal of it can be done with small telescopes. Furthermore, professional astronomers can be expected to steer clear of this field because it requires large amounts of telescope time without promising quick results.

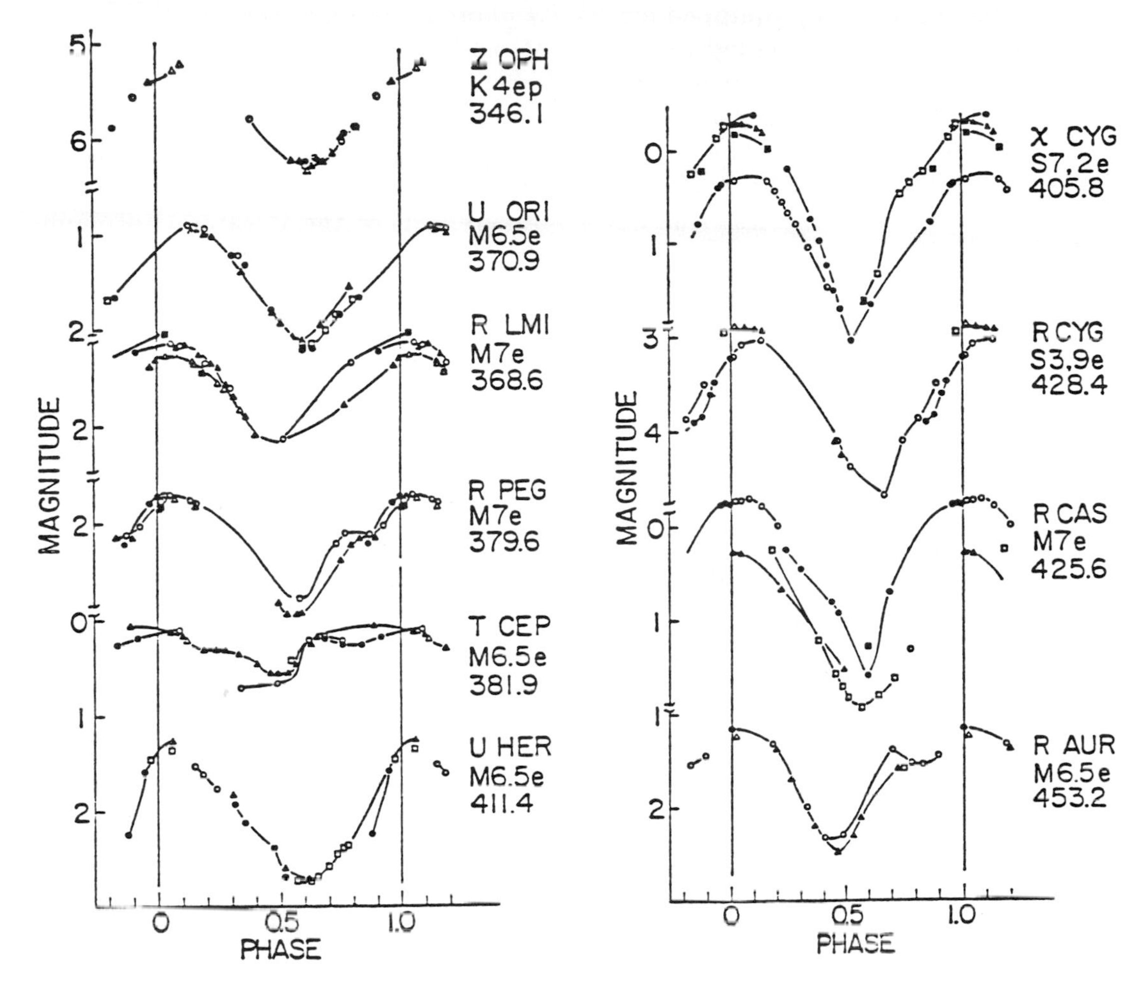

Figure 4. Light curves of ten Mira variables in I(104). Since this magnitude is unaffected by emission lines or molecular bands, these curves reveal the character of the bolometric variations. Note the humps on the rising branches of the light curves of several stars, and the large cycle-to-cycle differences. To the right of each curve is the star's name, its spectral type at a typical maximum, and the mean period which was used in calculating the phases. From Lockwood & Wing (1971).

The long-term monitoring of Mira variables in the infrared strikes me as an ideal project for a small college observatory. Students come and go, participating for varying lengths of time, but each one can contribute significantly to a data bank that becomes steadily more valuable.

Observers already equipped for $BVRI$ photometry may be wondering about the usefulness of measuring light curves of Miras in R and I. Such work certainly should not be discouraged, since very few light curves in R and I have been published. But it should be understood that, in terms of information content, the R and I curves are more like the V curve than the $I(104)$ or K curve. This is because, as Table 2 shows, the V, R, and I filters are all affected by TiO blanketing, although to different degrees, whereas a measurement of the infrared continuum really does provide additional information. If one knows the V and $I(104)$ curves one can quite accurately predict the R and I curves simply by interpolation, since the intermediate-wavelength filters also have intermediate amounts of blanketing by TiO. But I don't think one can safely predict the bolometric curve by extrapolating from $BVRI$ photometry; one needs a direct measure of the infrared continuum. Therefore, if a $BVRI$ photometer can be adapted for $I(104)$ photometry simply by purchasing an additional filter, I would recommend doing so. My own measurements of $I(104)$ have used rather narrow bandpasses (usually 50 Å), but the spectra of M stars in the 1 μm region are clean enough to allow bandwidths up to 300 or 400 Å if the central wavelength is shifted somewhat shortward to avoid the VO absorption that occurs in very cool stars (Persha 1986).

Finally, I should add a note about the effect of atmospheric water vapour on measurements made from sea-level observatories. In the case of $I(104)$ there is no problem, since the entire region from 1.00 to 1.08 μm is almost completely free of atmospheric absorption lines of any kind. The 2 μm region, however, is not as clean, and a K filter intended for use at low altitudes should be designed with that in mind. The standard K filter is centered at 2.2 μm and is very broad, extending from 2.0 to 2.4 μm and taking in the entire "window" between strong water bands. The transparency of the window is high at the centre but deteriorates toward the edges, primarily because of water lines. I would recommend using a filter only half as wide as the standard one, extending from 2.1 to 2.3 μm. Such a filter would give about 60% as much signal as the standard filter and would avoid problems due to variable water absorption almost completely.

III. MEASUREMENTS OF TiO STRENGTH BY NARROW-BAND PHOTOMETRY

Narrow-band photometry is a way of obtaining spectroscopic information by ordinary photometric techniques. A photometrist can become a spectroscopist simply by purchasing a set (or even just a pair) of narrow-band filters.

The spectral features best suited for measurement by narrow-band photometry are the strong, broad ones such as molecular bands. The most obvious examples are the TiO bands which dominate the spectra of M-type stars. Since the TiO bands are very sensitive to temperature (and since they are so pervasive that they make it hard to measure anything else), they are used to define the subdivisions of type M. Thus a narrow-band index of a TiO band, calibrated through observations of standard stars, can be used to obtain spectral types for M stars.

There are numerous ways to define a pair of filters which would give a TiO index in M stars. One filter should be in a heavily-depressed region, the other in as clear a continuum region as possible, and at the same time the filters should be close in wavelength to make the index insensitive to the slope of the star's energy distribution, which can be affected by extraneous factors such as interstellar reddening. If a third filter is added, an index can be defined that is completely independent of the slope.

Much of my research on cool stars has centered on photometry obtained with a set of eight narrow-band interference filters in the near infrared (0.7 to 1.1 μm). The filters are approximately 50 Å wide and were chosen to measure the strongest bands of TiO, VO, and CN as well as the best available continuum points in both M stars and carbon stars. I would like to describe this photometric system and some of its applications very briefly and then discuss a simplification of the system introduced by Persha (1986) to make it more suitable for use by amateur observers.

When eight-colour observations of program stars are reduced relative to standard stars (whose magnitudes on the eight-colour system have been expressed on a scale of absolute fluxes) the results can be plotted against wavelength to give an eight-colour "spectrum". In Figure 5 the spectra of typical stars of approximate types M2, M4, M6, and M8 are shown. Each one has been fitted with a blackbody continuum curve, the temperature of which is taken to be the "colour temperature" of the star. Filters 1 and 3 (in order of wavelength) are both depressed by TiO absorption, filters 4 and 8 by CN, and filter 6 by VO (but only in extremely cool stars). In carbon stars the oxide bands are absent while CN is very strong. White & Wing (1978) have discussed the procedures used to obtain spectral classifications from the photometry.

By using these filters at fairly large telescopes (the Perkins 72 inch, Kitt Peak's 50 inch, and Cerro Tololo's 60 inch) I have been able to classify some rather faint stars. These include more than 100 solar-neighbourhood M dwarfs, the M giants in a dozen old open and globular clusters, and M supergiants in the Large Magellanic Cloud. These classifications are two-dimensional (since the CN index indicates the luminosity class), and unlike conventional classifications, they include a colour index separately from the spectral type, so that the combination of the two can be used to determine the interstellar reddening.

Although one tends to observe faint stars if one can, there are literally thousands of red giants and supergiants bright enough to observe with telescopes much

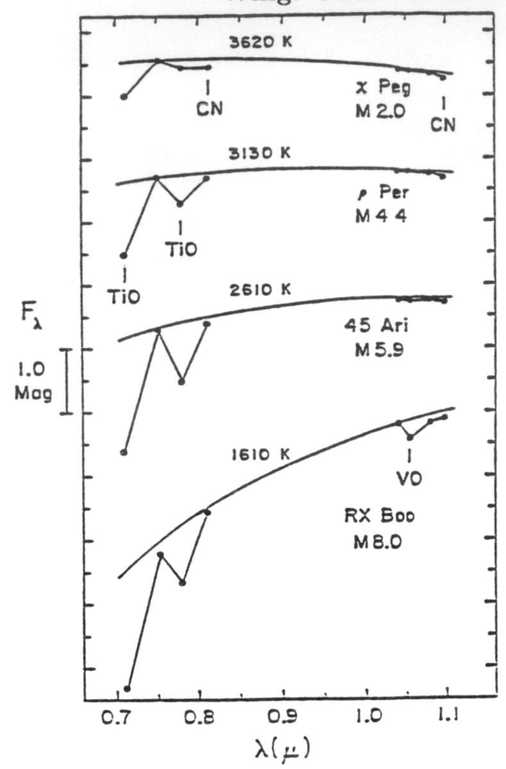

Figure 5. Spectra of four M stars as recorded on the eight-colour narrow-band photometric system. Absorptions due to the TiO, VO, and CN molecules are identified. Each set of photometry is accompanied by a blackbody curve representing the approximate location of the continuum and indicating the star's colour temperature.

smaller than the ones mentioned above. These include the 1000 or so red giants in the *Bright Star Catalogue* as well as nearly all of the 5000-plus stars recorded in the *Two Micron Sky Survey* or IRC. When the eight-colour system was new (1969-1971), I made a list of the 1000 brightest stars in the sky at a wavelength of 1 μm and observed all of them with the 16 inch telescopes at Kitt Peak and Cerro Tololo. Many amateurs have as much aperture as that.

The reason I would not expect amateur astronomers to use my eight-colour system is not that their telescopes are too small but that a set of eight interference filters is rather expensive (about $1000) and hard to mount in a small photometer. It is therefore worth considering whether a reduced version of the photometric system can still provide useful information.

From the representative spectra of Figure 5 we see that only three filters are needed to measure an infrared continuum magnitude, a colour index, and a TiO index. The best choice would be filters 1, 2, and 5 (centered at 7120, 7540, and 10400 Å) since filter 1 measures the strongest TiO band while filters 2 and 5 define the continuum. The magnitude at filter 5 is, in fact, what I've been calling the $I(104)$ magnitude.

Persha (1986) has acquired filters similar to filters 1, 2, and 5 of the eight-colour system, but with wider bandpasses, and has used them in an Optec SSP-3 photometer with a silicon photodiode detector. He finds that the wider bandpasses do not significantly reduce his ability to measure the quantities of interest while bringing more stars into the range of a small telescope.

Photometrists who tire of measuring light curves sometimes ask what they can do to add a new dimension to their program or to make their data more useful. To perform spectroscopy at each point on the light curve, while obviously desirable, is a tall order if it involves completely different equipment and long exposure times. But if one can measure spectral types simply by placing suitable filters in an existing photometer, one can easily define an interesting program of combined photometry and spectroscopy of red variables.

In the previous section we saw that the light curves of Mira variables in $I(104)$ are important because they are similar to the bolometric curves, and that they need to be monitored because they do not repeat well from cycle to cycle. With a filter set similar to the one described by Persha (1986), one can measure the $I(104)$ curves and at the same time determine the spectral type corresponding to each point on the light curve. This was done for 25 stars by Lockwood & Wing (1971) but this line of research has not been pursued.

Many Mira variables have spectral types of M8 or later, especially near minimum light, and for these stars it is useful to add a measurement of the VO band near 1.05 μm. This band is only 0.25 mag deep at type M8.0 (see Figure 5), but it grows rapidly with decreasing temperature and provides a good way to define the spectral types from M8 to M10. The TiO index given by filters 1 and 2 is an excellent spectral type indicator as far as M6 but thereafter it becomes less reliable because the continuum filter is no longer a good continuum point. Observers intending to emphasize the cooler Mira variables should therefore consider adding a fourth filter to measure VO.

I hope that these remarks will attract a number of variable-star observers into the area of infrared photometry of red variables, where there is so much work to be done. Interested observers are encouraged to write to me for information on standard stars, reduction techniques, and the like, and to Gerry Persha (Optec Inc., 199 Smith, Lowell MI 49331, USA) for advice on instrumentation.

REFERENCES

American Association of Variable Star Observers. (1983). *Observations of Long Period Variables*, 9 September 1974 - 5 June 1977, AAVSO Report **38**.

Campbell, L. (1955). *Studies of Long Period Variables,* Amer. Assoc. Variable Star Observers.

Lockwood, G.W. & Wing, R.F. (1971). *Ap. J.*, **169**, 63.

Persha, G. (1986). Contributed paper presented at this symposium.

Pettit, E. & Nicholson, S.B. (1933). *Ap. J.*, **78**, 320.

Smak, J.I. (1966). *Ann. Rev. Astr. Ap.*, **4**, 19.

INFRARED OBSERVATIONS OF CEPHEID VARIABLES

Douglas L. Welch
Department of Astronomy
University of Toronto
Toronto, Ontario
Canada M5S 1A1

Cepheid variables have long been considered the most trusted and reliable indicators of distances to nearby galaxies. In the last five years, it has been realized that there are major advantages to determining such distances from *near-infrared* observations rather than using the traditional B or V bandpasses. Specifically, observations made at $J(1.20\mu\text{m})$, $H(1.65\mu\text{m})$, and $K(2.20\mu\text{m})$ are much less affected by absorbing dust, blue main-sequence companions and atmospheric metallicity. Furthermore, the full amplitude of the light curve is significantly reduced in the near-infrared, resulting in a useful mean magnitude from one or a few points, unlike optical studies of Cepheids. The reader is directed to McGonegal *et al.* (1982) for an exposition of these advantages.

Since 1981, our group has been engaged in a recalibration of the local extragalactic distance scale using infrared photometry. Galactic calibrations of the infrared period-luminosity (P-L) relations have been published by McGonegal *et al.* (1983) and Welch *et al.* (1985a). Distances have been obtained to the Large Magellanic Cloud (LMC) and Small Magellanic Cloud (SMC) (McGonegal *et al.* 1982; Welch and Madore 1984; Welch *et al.* 1985b; Welch 1985), NGC 6822 (McAlary *et al.* 1983), IC 1613 (McAlary, Madore and Davis 1984), NGC 2403 (McAlary and Madore 1984), M33 (Madore *et al.* 1985), M31 (Welch *et al.* 1986), and NGC 300 (in preparation). These distance determinations have involved many of the largest telescopes in the world. Yet without observations from the 0.6 m telescope on Mt. Wilson, the interpretation of the large telescope data would be far less certain. It is the purpose of this paper to describe the importance of the small telescope observations and what has become of them in the intervening years.

During the summer of 1982, the 0.6 m reflector operated by the California Institute of Technology on Mt. Wilson was made available for JHK observations of Galactic Cepheids in a collaborative arrangement between Gerry Neugebauer and the Infrared Group at the University of Toronto. Fred Wieland (a senior at Caltech), Victoria Paylor, and I obtained photometry from June to September resulting in 523 JHK measurement sets of 25 Cepheids on 33 nights. The unusually long observing run resulted in full light curves for even the longest-period Cepheids. Such long-period Cepheids are representative of the extragalactic Cepheids typically selected for observation. Thus, full light curves for these stars are particularly valuable. Details regarding the observations may be found in Welch *et al.* (1984).

The Galactic light curves have since been used to correct single-phase observations of extragalactic Cepheids (made with large telescopes) to mean light. This correction results in the very well-defined LMC period-luminosity relation shown in Figure 1. The data illustrated in the figure have *not* been corrected for the effects of the relatively small range of reddenings present in the LMC. For comparison, the B period-luminosity relation is shown. The increased scatter at B is the result of the much greater sensitivity of the B bandpass to temperature and reddening differences.

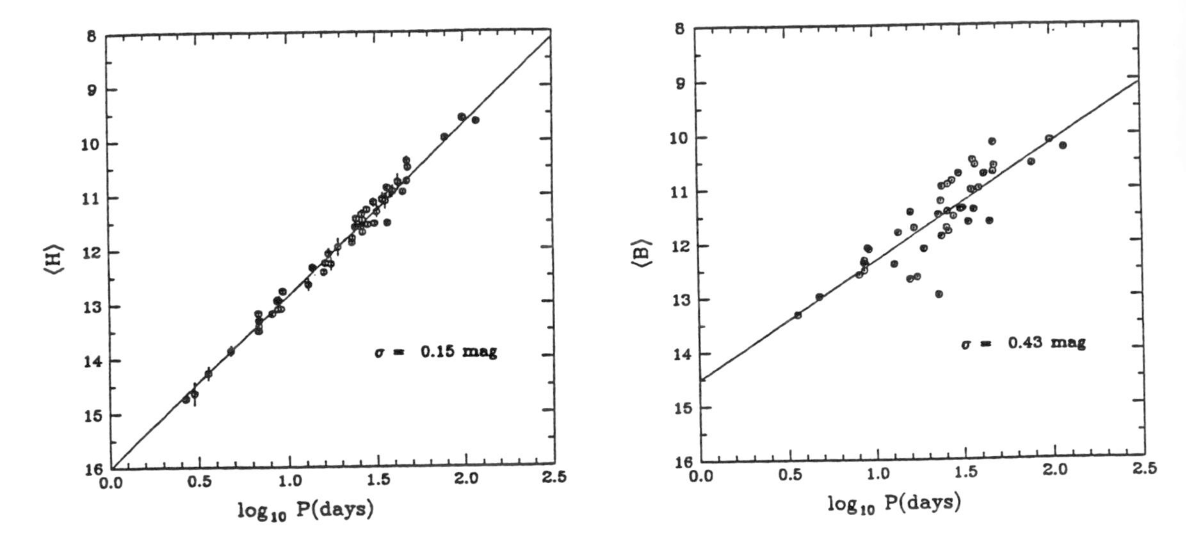

Figure 1. The mean H and B period-luminosity relations are shown here for 52 and 27 LMC Cepheids, respectively. The greater scatter at B is explained in the text.

The lack of temperature sensitivity at H has been put to good use in an entirely different project: the determination of Cepheid radii. Traditionally, Cepheid radii have been determined using the B and V bandpasses. Unfortunately, the change in temperature during the pulsation cycle of a Cepheid strongly dominates the B and V light curves, all but masking the flux variation caused by the change in stellar radius. By basing the radius determination on the K light curve, the effect of temperature change is greatly diminished resulting in better determined radii (given photometry of the same quality). Figure 2 illustrates the relative contributions of temperature (as manifested by the colour index $B - V$) and radius change to the V and K light curves, respectively. The K light curve is clearly more sensitive to the radius variations. This work has been carried out in collaboration with Dr. Nancy R. Evans (Computer Science Corporation) and Gordon Drukier (Univ. of Toronto).

There is yet work to be done on near-infrared Galactic Cepheid light curves. A number of bright Cepheids are in the winter sky and hence have no modern JHK light curves. The brightest of these are ζ Gem and T Mon. High-quality JHK photometry of these stars is eagerly awaited.

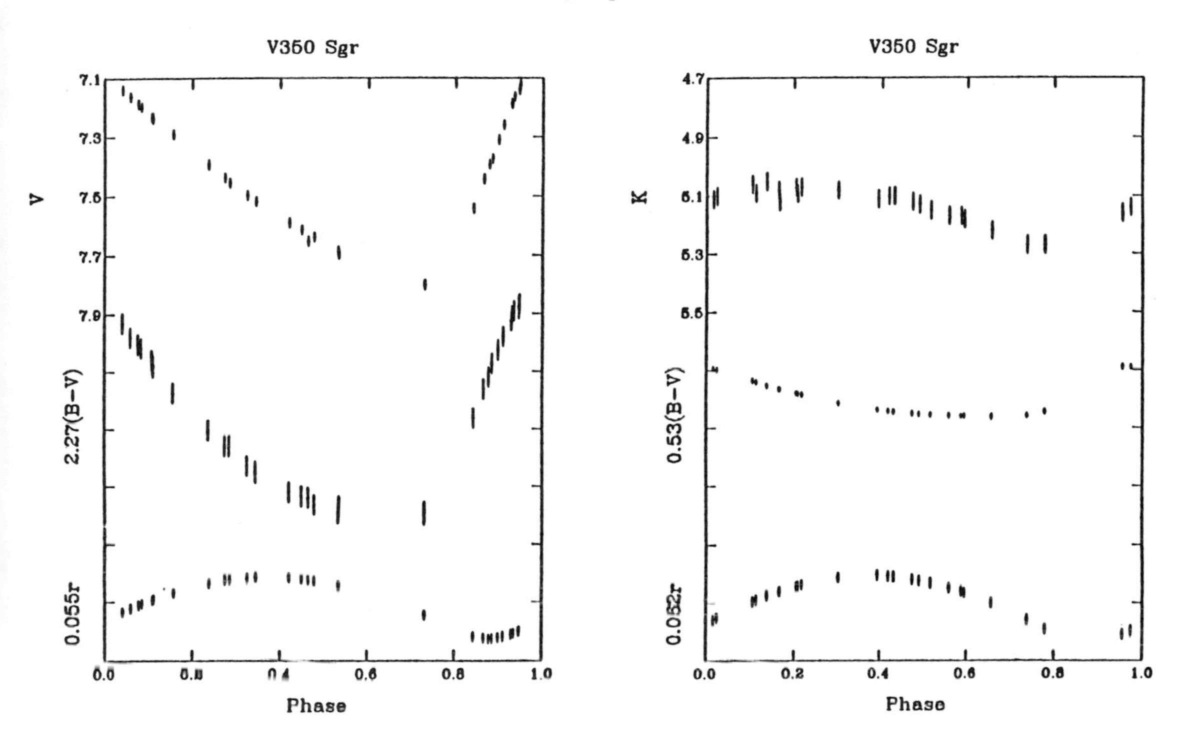

Figure 2. This figure illustrates the relative contributions of temperature and radius change to the V and K light curves of V350 Sgr. The upper curve in each figure is the light curve and the middle and lower curves show magnitude change caused by temperature and radius variation, respectively.

REFERENCES

Madore, B.F., McAlary, C.W., McLaren, R.A., Welch, D.L., Neugebauer, G., & Matthews, K. (1985). *Ap. J.*, **294**, 560. [M33]

McAlary, C.W. & Madore, B.F. (1984). *Ap. J.*, **282**, 101. [NGC 2403]
McAlary, C.W., Madore, B.F., & Davis, L.E. (1984). *Ap. J.*, **276**, 487. [IC 1613]
McAlary, C.W., Madore, B.F., McGonegal, R., McLaren, R.A., & Welch, D.L. (1983). *Ap. J.*, **273**, 539. [NGC 6822]
McGonegal, R., McAlary, C.W., Madore, B.F., & McLaren, R.A. (1983). *Ap. J.*, **269**, 641. [Preliminary Galactic calibration]
McGonegal, R., McAlary, C.W., McLaren, R.A., & Madore, B.F. (1982). *Ap. J. (Letters)*, **257**, L33. [LMC demonstration]
Welch, D.L., McAlary, C.W., Madore, B.F., McLaren, R.A., & Neugebauer, G. (1985a). *Ap. J.*, **292**, 217. [Galactic calibration]
Welch, D.L., McAlary, C.W., McLaren, R.A., & Madore, B.F. (1985b). In *Cepheids: Theory and Observations*, IAU Colloquium 82, ed. B.F. Madore, p. 219. Cambridge: Cambridge University Press. [LMC and SMC]
Welch, D.L. & Madore, B.F. (1984). In *Structure and Evolution of the Magellanic Clouds*, IAU Symposium 108, ed. S. van den Bergh & K.S. de Boer, p. 221. Dordrecht: D. Reidel. [LMC and SMC]
Welch, D.L., Wieland, F., McAlary, C.W., McGonegal, R., Madore, B.F., McLaren, R.A., & Neugebauer, G. (1984). *Ap. J. Suppl.*, **54**, 547. [Galactic Cepheids]
Welch, D.L., McAlary, C.W., McLaren, R.A., & Madore, B.F. (1986). *Ap. J.*, (submitted). [M31]
Welch, D.L. (1985). *Ph. D. Thesis*, University of Toronto.

VARIABLE STAR RESEARCH AT THE UNIVERSITY OF TORONTO SOUTHERN OBSERVATORY (UTSO) AT LAS CAMPANAS, CHILE

Robert F. Garrison
David Dunlap Observatory
University of Toronto
Toronto, Ontario
Canada M5S 1A1

ABSTRACT

In the Atacama Desert, the skies are extraordinarily clear, there are no city lights within a hundred miles, and the seeing is excellent. Unlike most other small telescopes, the UTSO 0.6 m is equipped with high-quality instrumentation, allowing sophisticated spectroscopic, photometric, and imaging programs to be carried out. Because the telescope is small, time assignment is not as limited as with large telescopes at good sites, and it is possible to carry out observing programs over weeks, months, or years. With all these advantages, the UTSO is unique among small telescopes for the study of variable stars.

Extensive research on long-period and cepheid variable stars has been carried out with the UTSO 0.6 m telescope. At the present time, a long-term Hα survey of T Tauri stars is being carried out by the Resident Observer for Bill Herbst of Wesleyan University. Christine Clement reports elsewhere in these proceedings on some aspects of her long-term work with globular cluster variables.

THE SITE*

It may seem odd that Canadian astronomers go to Chile to study the stars, but there are two good reasons. First and foremost, the southern sky contains many scientifically exciting and unique objects which have long been neglected. Secondly, the Canadian climate, as well as the rapid growth of Canadian communities like Toronto, Victoria, and London is rendering nearby telescopes less effective than they used to be for some (though not all) types of research.

In Chile, on mountain peaks lying between the high Andes and the Pacific Ocean at elevations of 2000 meters or so, conditions for astronomical observations

---

* This brief description of the site is based on a more detailed one by MacRae & Hogg (1973).

are as nearly perfect as they can be on this Earth. The University of Toronto's 0.6 m (24 inch) telescope is located on Las Campanas at the southern edge of the Atacama Desert. Through the cooperation of the Carnegie Institution of Washington, the University of Toronto was able to make arrangements to have a share in this site. The Carnegie Southern Observatory (CARSO) operates the 1 m (40 inch) Swope reflector and the 2.5 m (100 inch) Dupont telescope on Las Campanas, which is at a latitude of 29° south and a longitude of 71° west.

There are two ways to reach Las Campanas. It is possible to fly in a light plane directly to an airstrip operated by the European Southern Observatory at the bottom of the mountain. Alternatively, a seven-hour express bus trip from Santiago takes the observer to La Serena. From there, a Carnegie vehicle makes a trip of two hours to the mountain top. It goes north about 130 km on the Pan-American Highway, then turns towards the mountains for 50 km along a dirt road. Las Campanas is actually a narrow ridge running for 7 km in a northwest-southeast direction. The UTSO 0.6 m telescope is located at an altitude of 2250 m, near the northern end of the ridge.

THE TELESCOPE AND ACCESSORY EQUIPMENT

The 0.6 m telescope was purchased from Ealing in Montreal, but was built in Cambridge, Massachusetts by Competition Associates. The Cassegrain setup works at f/15. When used for direct photography, the telescope is fitted with a camera which takes plates 10 × 13 cm (4 × 5 inches) in size. The scale of these plates, 22.54 arcseconds/mm, is very close to that of the Newtonian focus of the 1.88 m (74 inch) reflector at Richmond Hill. An ITT magnetically-focussed image-tube camera, which has a useful field of about 12 arcminutes, may also be used.

A grating spectrograph designed by the author has seen a great deal of use. It may be employed with photographic plates, directly or with a one-stage image-tube interposed, at dispersions of 120 Å/mm and 67 Å/mm. Three photon-counting photoelectric photometers are also available. Photomultipliers on hand are of three types: S4 for UBV wavelengths, S25 extended red for RI, and Indium-Antimonide for the infrared JHK.

There is an on-line reduction system complete with two Osborne-1 micro-computers, with an HP-9815A and an HP-85 as backup. By the end of this year, an echelle spectrograph with a Reticon detector will be in use, and two CCD detector systems have been applied for. The scale and availability of this small telescope make it ideal for survey programs, which are difficult, if not impossible for large telescopes.

Other auxiliary instrumentation is available to fit particular astronomical programs. To facilitate operations, a short-wave radio is available for twice-weekly (or more) communications with the DDO. There is also a large tape collection (over 100 7 inch reels) to provide musical comfort for the observers.

THE SERVICE BUILDING

Besides the dome, which houses the telescope, the University has built a small combination residence and operations centre, constructed of local stone so as to blend into the mountain slope. One full wall of windows overlooks the plains and foothills dropping westward to the Pacific Ocean. The house is located just off the ridge to be away from the strong winds. The building provides comfortable accommodation for two observers, desk and work space, facilities for light meals, and a fully-equipped dark room.

THE RESIDENT OBSERVER

It is now the established procedure to have one resident astronomer who maintains the equipment and carries out routine observing for absentee guest investigators, thus saving the expense of a trip when just a few observations are needed. Since August, 1971, when operations began, there have been eleven resident observers. These individuals typically have bachelor's or master's degrees in astronomy, along with practical skills in optics and electronics. Several have later moved on to positions at larger observatories, or have returned to university to pursue a Ph.D. degree in astronomy.

THE SOUTHERN SKIES

Understandably, research programs are closely related to the special virtues of the southern skies. If it is autumn in the Southern Hemisphere (March-June), the centre of our Milky Way Galaxy will be directly overhead late into the night. During their spring and summer, the Magellanic Clouds will stand out not many degrees from the South Celestial Pole. It is an awesome moment when, standing on Las Campanas under a black, moon-less, clear sky, you are enveloped by the brilliant semicircle of the Milky Way, from horizon to horizon!

VARIABLE STAR RESEARCH

There are many special advantages to having such a telescope at a good location. One of them is that it allows us to carry out variable star surveys, which are impossible with large telescopes or cloudy skies.

Two large spectroscopic surveys, a large photographic survey (described elsewhere by Christine Clement), and several photoelectric surveys have been carried out. In addition, several new variables have been discovered serendipitously during surveys of various kinds.

THE "PEOPLE'S CATACLYSMIC VARIABLE"

The most spectacular example of serendipity is that of CPD $-48°1577$, which is the brightest cataclysmic variable known. Many faint ones are known, but these can only be studied with large telescopes: hence the name "people's cataclysmic variable".

It was discovered during the MK spectral classification survey of southern OB stars carried out by Garrison, Hiltner & Schild (1977). Only a hot continuum is visible at first glance, but on careful inspection, it is possible to see some very broad, shallow hydrogen lines with emission cores. Observations in *UBV* confirmed the suspicion that it is similar to the well-known cataclysmic variable star UX UMa. At a magnitude of 9.4 (varying from 9.4 to 9.9), CPD $-48°1577$ is the brightest of all the known cataclysmic variables.

The currently popular CV model is a binary system in which a cool star transfers matter onto an accretion disk surrounding a white dwarf companion. Most of the observed optical and x-ray emission is presumed to originate in the accretion disk. A hot spot where the stream of matter strikes the accretion disk contributes significantly to the emission. The spectrum is characterized by very broad, shallow hydrogen absorption lines and emission cores with a very slow Balmer decrement, quite unlike that of a normal Be star. Also present are extremely weak absorption lines of He I and He II. He II 4686 is faintly visible in emission, and some of the other He II lines may have emission components, but they are too faint to be certain. These characteristics are consistent with those expected for the standard model of a cataclysmic variable (Garrison, Hiltner & Krzeminski 1984).

SPECTROSCOPY OF CEPHEIDS

Gauthier studied the spectra of 26 Southern Hemisphere Cepheids around their cycles. His is the most comprehensive classification study ever carried out for these fascinating stars. His results are published in his 1983 doctoral thesis and in an article in the book *The MK Process and Spectral Classification* (Garrison 1984).

SPECTROSCOPY OF MIRA VARIABLES

Crowe extended the catalog of Mira variables of Keenan, Garrison & Deutsch (1974) to include the southern Miras. Using direct and image tube plates, he took 475 spectra of 72 variables around their cycles, at a dispersion of 120 Å/mm. This work represents a very considerable effort over about 6 years. Many of the stars at minimum were quite faint for such a small telescope. His results are published in his 1983 doctoral thesis and the *Journal of the RASC* (Crowe 1984) as well as the *Journal of the AAVSO* (Crowe 1983) and the book *The MK Process and Spectral Classification*, edited by Garrison. The complete version will be published as an *Astrophysical Journal Supplement* in 1986.

T TAURI STARS: A LONG LOOK

A program currently underway is being carried out by the Resident Observer for Bill Herbst of Wesleyan University. It consists of photoelectric observations of Hα in southern T Tauri stars. These fascinating stars vary irregularly, and the plan is to monitor them continuously for several days of each month over a period of years. This is the kind of work that is impossible to do with a large telescope, with inflexible scheduling. Yet, there is a great need for this kind of astronomy, as a complement to the short-term projects carried out with large telescopes.

SUMMARY

Programs such as these can only be done at an excellent site with a small telescope, because larger telescopes are too much in demand. Good weather and dark skies are important for good phase coverage. In Chile, the excellent seeing, clear skies and freedom from city lights help to make observing with a small telescope very effective in producing good scientific results.

REFERENCES

Crowe, R.A. (1983). *J. Amer. Assoc. Var. Star Obs.*, **12**, 58.
Crowe, R.A. (1984). *J. R. A. S. Canada*, **78**, 103.
Garrison, R.F. (1984). *The MK Process and Spectral Classification*, Toronto: David Dunlap Observatory.
Garrison, R.F., Hiltner, W.A., & Schild, R.E. (1977). *Ap. J. Suppl.*, **35**, 111.
Garrison, R.F., Schild, R.E., Hiltner, W.A., & Krzeminski, W. (1984). *Ap. J. (Letters)*, **276**, L13.
Keenan, P.C., Garrison, R.F., & Deutsch, A.J. (1974). *Ap. J. Suppl.*, **28**, 271.
MacRae, D.A. & Hogg, H.S. (1973). *Physics in Canada*, **32**, 23.

ON THE SPECTRAL VARIABILITY OF THE EXTREME Of STAR, HD151804

S. Jeffers
Physics Dept.
York University
4700 Keele St.
Downsview, Ontario
Canada M3J 1A3

W.G. Weller
Cerro Tololo Interamerican Observatory
Casilla 603
La Serena, Chile

GENERAL REMARKS

The symposium was largely concerned with photometric techniques for small telescopes with some discussion of photographic spectroscopy. This paper discusses spectrophotometric observations made with a sensitive multi-channel detector. Such detectors can be used on small telescopes (i) to obtain spectra of objects too faint to study using photographic recording; (ii) to make repeated observations at short time intervals for rapid variability studies; (iii) to do *spectrophotometry* at moderate resolution by obtaining extinction data and determinations of the instrumental spectral response. In a three-week observing run on the 60 cm telescope of the University of Toronto at Las Campanas, one of us (W.G.W.) obtained over 2000 spectra of various objects including Wolf-Rayet stars, Of stars and X-ray binaries. The data on WR stars were used by W.G.W. for his thesis. The data from this one observing run have generated a number of publications (most recently, Jeffers & Weller 1985a; Jeffers, Stiff, & Weller 1985b) with more in preparation. This paper illustrates the kind of work that can be done with a well-equipped small telescope in a good location.

INTRODUCTION

HD151804 is an extreme Of star ($m_V = 5.22$, O8Ifp) and a member of the Sco OB1 association. Its absolute visual magnitude is -7.2 (Hutchings 1976). The spectrum of this star has been described in detail by Hutchings (1968) who measured the velocities of many lines from high dispersion plates. Velocities measured for different ions, hydrogen and for He I show that the atmosphere of this star is extended and accelerating. The density of material in the accelerating atmosphere is high enough to produce He II (468.6 nm) in emission. Variability in Of stars has been studied by a number of authors (see Table 1). Early reports (Brucato 1971) of variability on a time scale of minutes have not been substantiated (Lacy 1977; Weller & Jeffers 1979). Recently, Grady *et al.* (1983) have studied at high dispersion λ468.6 nm He II profile variations in four Of stars including HD151804. Night-to-night variations in the profile shape are reported. Here we report on low-dispersion spectroscopy covering the spectral range 375 – 500 nm. Time series ($\Delta t = 2$ mins) were obtained on three successive nights.

OBSERVATIONS

The data presented here were obtained with a low dispersion spectrograph (11.2 nm/mm in second order, $\Delta\lambda = 0.8$nm) equipped with an intensified silicon vidicon detector (Jeffers & Weller 1976; Weller 1984). The instrument was used on the 60 cm telescope operated by the University of Toronto at Las Campanas, Chile. HD151804 was observed on three successive nights (May 29/30, May 30/31, May 31/June 1, 1974). On each night a time series of 8 – 20 spectra were recorded with a temporal resolution of 2 minutes. The integration time for each spectrum was 19.8 seconds. Sky and instrumental backgrounds were recorded at the beginning and end of each time series. Comparable time series of HD170040 ($m_V = 6.65$, B8) were also obtained for comparison. Sky plus instrumental backgrounds were subtracted from each spectrum and the spectra from each time series were averaged. The mean spectra for each night are shown in Figures 1 to 3. For these averaged spectra, the continua were fitted to a polynomial and divided into the spectra. For each set of observations, the mean spectrum was subtracted from each spectrum in the time series. Mean and differenced spectra for J.D. 2442198.77 are shown in Figure 4 for the spectral region 454 – 478 nm.

DISCUSSION

The data of Figure 4 show no evidence of variability on a time scale of 2 – 40 mins. However, variability is seen in a number of spectral features from night to night. Table 2 shows measurements of the equivalent widths of the strong spectral lines for the three nights. Measurements of Hγ for the standard star HD170040 were used to establish the accuracy of the equivalent width measurements at $\pm$13%. Hγ weakens considerably on the third night with no comparable change in Hβ. P Cygni profile is seen at λ447.1 nm He I on all three nights. Hδ and 419.9 nm + 420.2 nm (He I + N III) develop strong P Cygni profiles on the second night. The emission at 463.4 nm (N III) remains constant but He II 468.6 nm weakens by a factor of five on the last night. Each spectrum in the time series for each night shows the same spectral characteristics, which indicates that the spectral variability is real

Table 1. Rapid line variability studies of Of stars.

Authors	# of Stars Observed	Observing Technique	Wavelength Observed	Time Resolution	Variability
Brucato 1971	6	Photographic & image tube spectroscopy	$H\beta - H\gamma$	10 mins	Yes – on time scale of minutes
Conti & Frost 1974	1	Coude spectroscopy	λ468.6, HeII	1 day	Yes – on time scale $<$ 1 day
Leep & Conti 1979	1	Coude spectroscopy,	λ468.6	1 day	Yes – night-to-night
		Copernicus Satellite	CIII 117.5 NV 123.8 NV 124.2		No
Snow, Wedner & Kunasz 1980	1		$H\alpha$ HeII 468.6	20 mins	Yes $\sim$ 1 hour
			SiIV 140.2 OVI 103.7	$\sim$ 40 mins	Yes $\sim$ few hours
Weller & Jeffers 1979	2	SIT Vidicon Spectroscopy	λ468.6, HeII $H\gamma$	2 mins	No – on time scale $<$ 1 hour
Grady, Snow & Timothy 1983	4	Coude spectroscopy	λ468.6, HeII	1 day	Yes – on time scale of 1 day

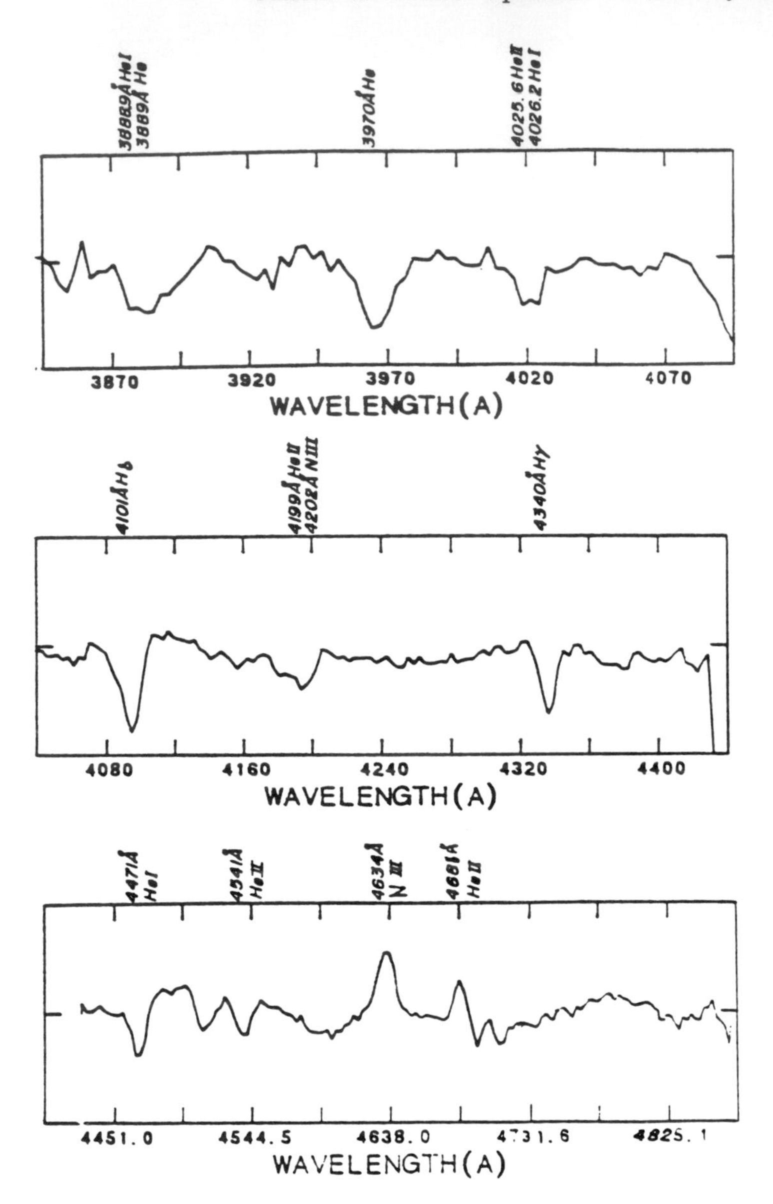

Figure 1. Mean of 8 spectra of HD151804 recorded on J.D. 2442196.84. The spectra have been reduced to continuum = 1.

and not due to some spurious instrumental effect. This conclusion is confirmed by inspection of the standard star data for each night.

Variability in the atmospheres of Of stars may arise due to a number of causes (Underhill 1983; Grady *et al.* 1983). Periodic variability in line profiles and equivalent widths could arise due to (i) asymmetric mass flow in a binary system

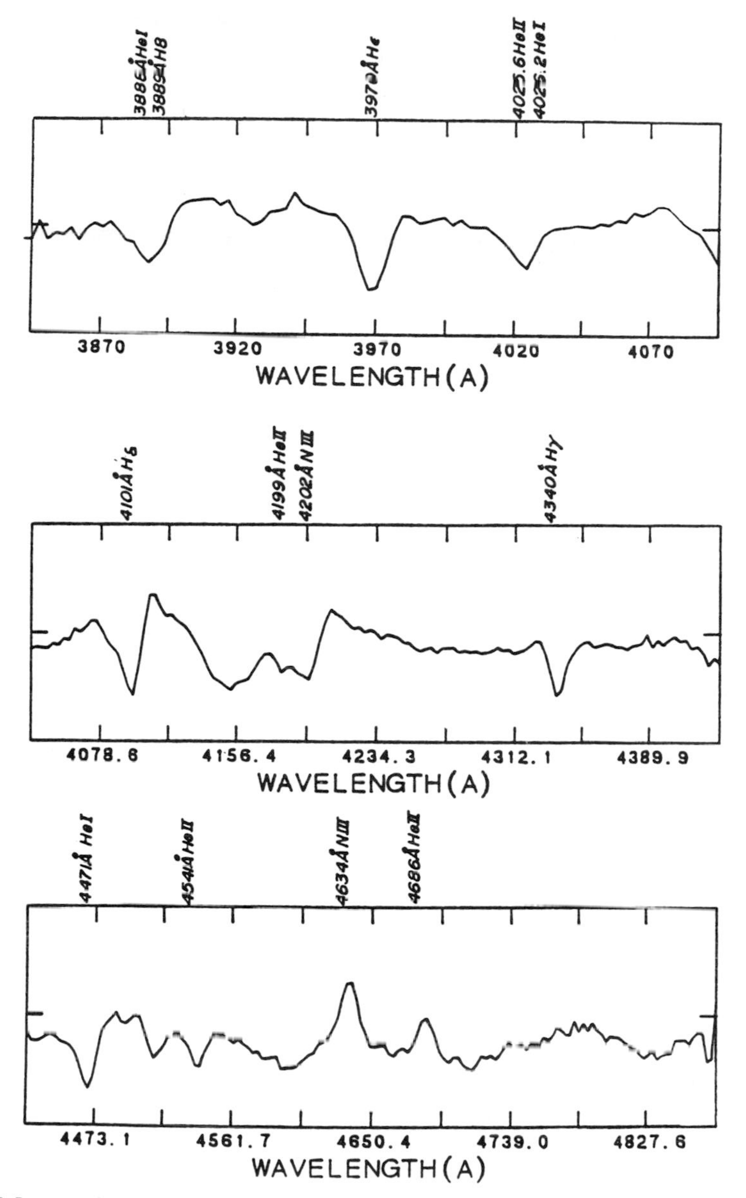

Figure 2. Mean of 8 spectra of HD151804 recorded on J.D. 2442197.67. The spectra have been reduced to continuum = 1.

(however, a radial velocity study by Conti *et al.* (1977) appears to establish that HD151804 is a single star), (ii) pulsation, (iii) asymmetries in the atmospheric conditions which are long-lived compared to the rotation period. Among the Of stars, HD151804 is a relatively slow rotator ($v \sin i = 50$ km sec^{-1}, Conti & Ebbets 1977). Grady *et al.* , assuming $i = 90°$, $R_* = 10^{12}$ cm, gives a rotation period of 34 days. This is a maximum figure. Given the arbitrariness of the assumption that

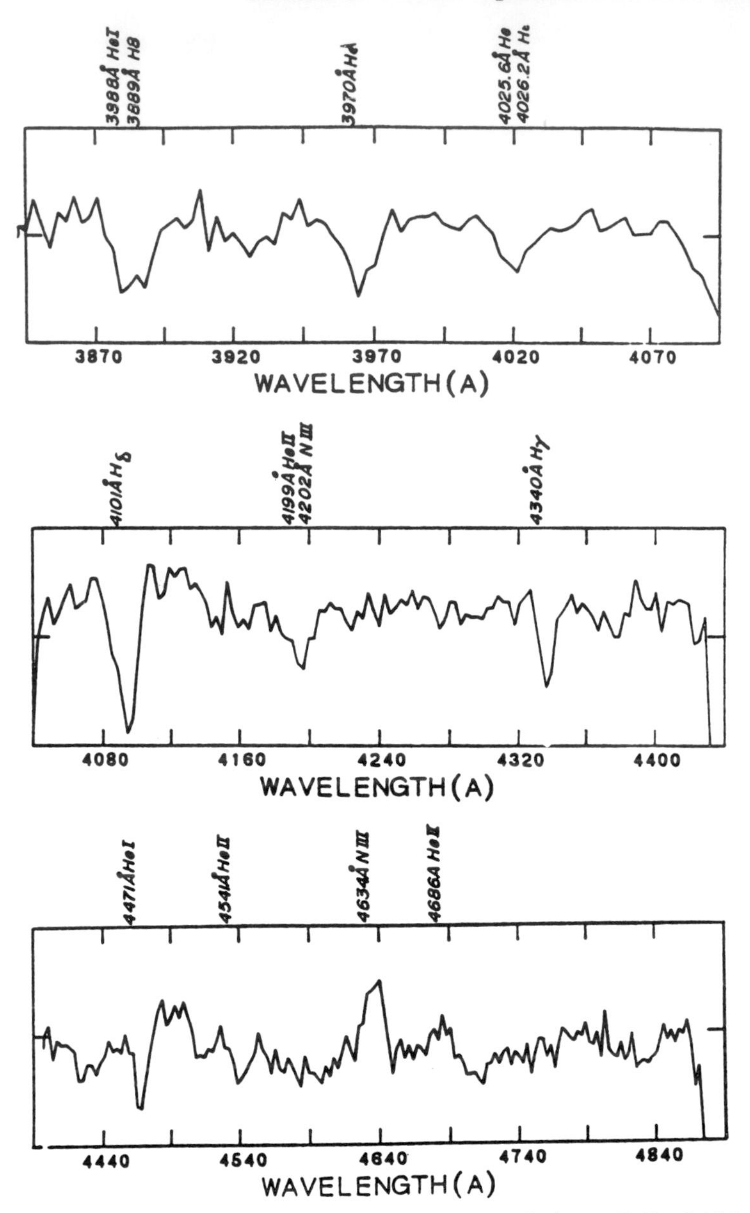

Figure 3. Mean of 20 spectra of HD151804 recorded on J.D. 2442198.77. The spectra have been reduced to continuum = 1.

$i = 90°$, neither the data presented here or that of Grady *et al.* (1983) can entirely rule out rotation as a significant factor in accounting for the observed variability. The data presented here show that the time scale of the variability is from ~ 1 hour to less than 1 day. This would tend to indicate a random perturbation in the physical conditions in the atmosphere (*i.e.*, variable mass flow or variation in the velocity structure of the wind) as the cause of the observed variability.

Table 2.

λ (nm)	I.D.	Equivalent Width (nm) J.D. 244 2196.84	2197.67	2198.77	Comments
388.965 388.904	He I H8	.132	.149	.105	
397.0	$H\epsilon$	.107	.140	.085	
410.1	$H\delta$	.104	.097	.127	P Cygni profile on J.D. 2442197.67
419.9 420.2	He N III	—	—	—	P Cygni profile on J.D. 2442197.67
434.0	$H\gamma$	.102	.081	.0334	Significantly weaker on J.D. 2442198.77
447.1	He I	.031	.031	.052	P Cygni profile on all three nights
454.1	He II	.062	.043	.048	
463.4	N III	−.234	−.249	−.235	Constant
468.6	He II	−.041	−.048	−.01	Weaker on J.D. 2442198.77
486.1	$H\beta$	.025	.042	.038	

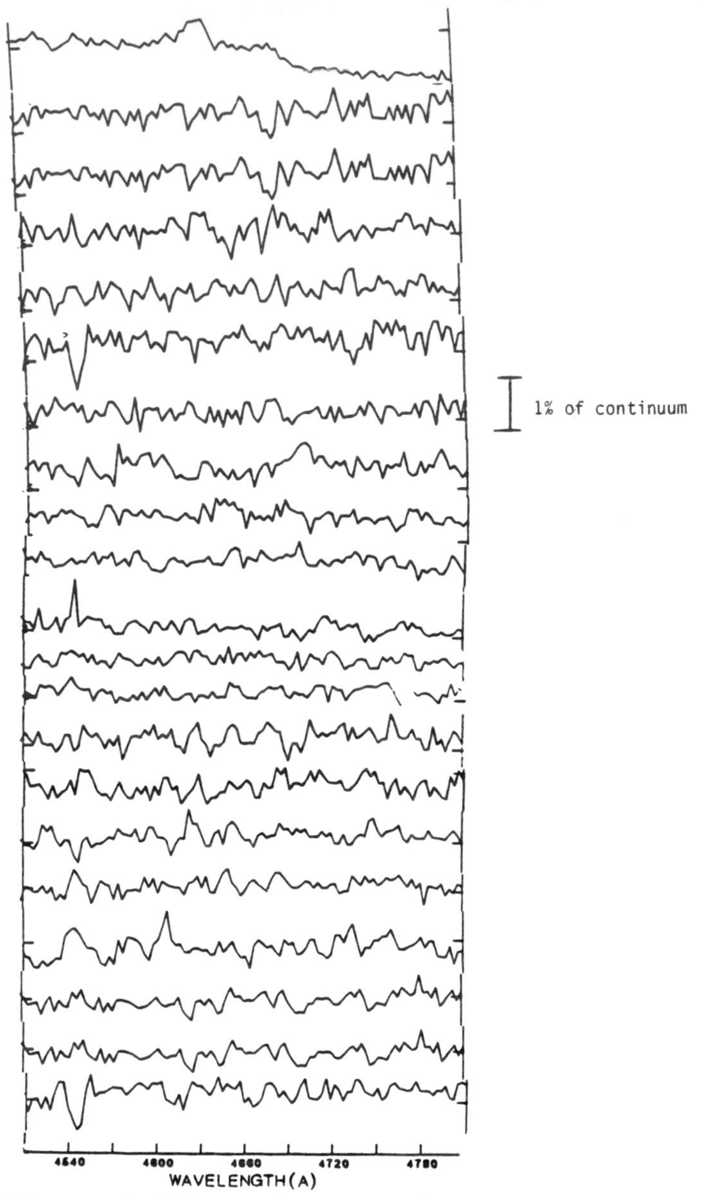

Figure 4. Mean and differenced spectra of HD151804 recorded on J.D. 2442198.77. Time interval between spectra = 2 minutes.

CONCLUSIONS

Spectral variability is observed to occur in HD151804 on a time scale of from 1 hour to less than one day. Given that HD151804 is a relatively slow rotator and a single star, this would indicate that the variability is the result of random perturbations in the stellar atmosphere.

Acknowledgements

S.J. gratefully acknowledges operating grant support from the Natural Sciences and Engineering Research Council of Canada. We would also like to acknowledge Mr. R. Jain, NSERC Summer Student at York University, for his assistance with data processing. We thank Prof. D.A. MacRae (former Director of the David Dunlap Observatory) and colleagues at the University of Toronto for their permission to use the Las Campanas Telescope.

REFERENCES

Brucato, R.J. (1971). *M. N. R. A. S.*, **153**, 435.
Conti, P.S. & Frost, S.A. (1974). *Ap. J. (Letters)*, **190**, L137.
Grady, C.A., Snow, T.P., & Timothy, J.G. (1983). *Ap. J.*, **271**, 691.
Hutchings, J.B. (1968). *M. N. R. A. S.*, **141**, 219.
Hutchings, J.B. (1976). *Pub. D. A. O.*, **14**, 355.
Jeffers, S. & Weller, W.G. (1976). *Advances in Electronics and Electron Physics*, Vol. 40B, p. 887. London: Academic Press.
Jeffers, S. & Weller, W.G. (1985a). *Astr. Ap.*, in press.
Jeffers, S., Stiff, T., & Weller, W.G. (1985b). *A. J.*, in press.
Lacy, C. (1977). *Ap. J.*, **212**, 132.
Leep, E.M. & Conti, P.S. (1979). *Ap. J.*, **228**, 224.
Snow, T.P., Wegner, G.A., & Kunasz, P.B. (1980). *Ap. J.*, **238**, 643.
Underhill, A. (1983). *Workshop on Rapid Variability of Early Type Stars*, Hvar Observatory Bulletin, Vol. 7, No. 1, p. 1, ed. P. Harmanec & K. Pavlovski.
Weller, W.G. & Jeffers, S. (1979). *Mass Loss and Evolution of O-type Stars*, ed. P.S. Conti & C.W.H. de Loore. Dordrecht: Reidel.
Weller, W.G. (1984). *Ph.D. Thesis*, York University. Toronto, Ontario.

V. COORDINATION AND ARCHIVING OF OBSERVATIONS

These two topics, which are both important in themselves, are interrelated in several ways. If only one observer collects data on a specific star, then that observer would most likely be the one to use and publish them. If many observers collect data, then they would normally send them to a particular individual or organization. It is the duty of that individual or organization to make the observations maximally useful and available. This, however, may require considerable effort. Even in the best of cases, data from multiple observers is usually less homogeneous. Chris Sterken's paper contains instructions for maximizing the value of the data, and minimizing the strain on the coordinator.

Chris Sterken has been active as both a photometrist and an organizer of coordinated observing campaigns; his presence also helped to give this meeting an international flavour. His philosophical comments about the nature and role of the amateur were not specifically requested, but came as a pleasant bonus. Chris Stagg and I have contributed our comments about two campaigns in which we have been active; we stress both the advantages and the problems of such campaigns.

Elizabeth O. Waagen, scientific assistant to the Director of the AAVSO, has special responsibility for implementing the computer-based archiving system at the AAVSO headquarters. Her paper provides an interesting case study and also a historical perspective on the growth of the AAVSO. From her position, she sees the problems of archiving from both a scientific and a technical point of view. The number of visual observations archived by the AAVSO each year is growing rapidly, and the problem of dealing with these while maintaining good quality control and accessability, is not an easy one. Archiving is certainly one of those processes which requires a great deal of forethought, and looks much easier than it actually is.

VARIABLE STARS AS TARGETS FOR LONG-TERM COORDINATED MULTISITE PHOTOMETRY

C. Sterken
Astrophysical Institute
University of Brussels
Pleinlaan 2
B-1050 Brussels
Belgium

In the course of the last few years I have collaborated with many professional colleagues at different observatories during observing campaigns of short and long duration. In 1979, on the occasion of my participation at a conference in New Zealand, I met some Australian and New Zealand amateur astronomers, and I was greatly impressed when I saw their achievements, which not only reflected a high level of observing skills, but also reliable and truly professional instrumentation. We were planning a worldwide observing campaign on EX Hydrae, a southern hemisphere cataclysmic variable, and we were very happy that some of these observers enthusiastically joined our team. So, one year later, professional observers in Chile and in South Africa, and amateur astronomers in Australia and New Zealand observed EX Hydrae almost continuously during more than two weeks. This campaign yielded very interesting results (Sterken *et al.* 1983) which could never have been obtained by a single observer at a single site.

Soon after this experience one of my colleagues, surprised that fruitful collaboration between professionals and non-professionals was possible, drew my attention to a statement taken from Mulholland (1982) and quoted by Williams (1983):

> "Today an amateur astronomer is one who does trivial astronomy in his spare time and with equipment that is normally not suitable for serious research."

Since I was unable to find the original paper from which this sentence was quoted, the statement gave me the impression of being a (short-sighted) definition of the term "amateur". What is an amateur? There certainly exist elaborate studies which precisely describe the term (see *e.g.*, Stebbins 1981, 1982), but in order to understand the evidence from which Mulholland's statement was derived, I tried to see the basic difference between amateur and professional activities. Therefore I browsed through the telephone directory yellow pages, making a classification of professions and trades in two categories: the first category contained all professions for which amateurs and even amateur associations exist; in group two were the professions where amateur activities do not exist. The first sample contained professions such as musician, poet, writer, actor, scientist, photographer, pilot, cook, gardener, etc.; the second list contained pharmacist, butcher, notary, dentist, banker, and so on. Although people having professions belonging to the second category may love their profession to a high degree, the category one professions clearly have the common aspect that they are activities which one pursues for amusement and for pleasure. This is not surprising, since the term "amateur" comes from its Latin root which means "lover".

The Oxford English Dictionary tells us that an amateur is one who cultivates anything as a pastime, but the French Larousse gives an extra characteristic: "Celui qui cultive un art ou une science pour son *plaisir* et avec *compétence* ..." The last description can very well be applied in the case of the amateur astronomer: doing science for one's own pleasure and in a competent way.

The term amateur is often confused with amateurish: the amateurish approach to any performance. We have all seen at some time amateurish achievements by professionals in *all* professions of both categories described above. More specifically, many a professional colleague will agree that sometimes professional achievements are of such a doubtful level that, by changing only a few words in Mulholland's sentence, one may write a similar statement concerning some professionals.

But undoubtedly Mulholland's statement is applicable in some cases, and he certainly has strong reasons for advancing three crucial points, *viz.*, (i) the present situation ("Today, ..."), (ii) the instrumentation, and (iii) the trivial work. Each aspect deserves some special attention.

First, one may think that the situation was better in the past. Indeed, there was a strong tradition of amateur contribution to astronomy, *e.g.*, famous comet hunters like J. Pons and C. Herschel, the discovery of the sunspot cycle by Schwabe, and the discovery of the period of Algol and the scientific hypothesis of binary motion, by John Goodricke.

But there is no reason to believe that today the situation has deteriorated. On the contrary, we all know that a large number of comets are still discovered by amateurs. Also, excellent work is done in the field of variable star work (*e.g.*, by members of AAVSO, GEOS, etc.).

The second argument perhaps comes from the impression that due to the increasing sophistication of professional instruments and working methods, amateur contributions are less valuable than they were in the past. However, while this is only true in the case of the largest telescopes equipped with high technology super expensive detecting devices, the overall availability of budget-priced smaller instruments and computer facilities have recently allowed several amateurs to contribute significantly to scientific research (Hall & Genet 1983). Moreover, the relative importance of the amateur contributions to the results obtained by professionals will even increase if only the amateur-professional communication and collaboration could be improved.

The last point, concerning the trivial work, is only valid if it refers to observations done for pure amusement (*e.g.*, deep-sky photographic work yielding high quality — but uncalibrated — images, which have purely esthetic value). In my personal opinion, work is never trivial if:

1) there is a good *scientific motivation* for carrying out specific observations;

2) one *measures*;

3) one measures *accurately*;

4) one takes *long strings of data measurements*;

5) one gets well *organized* and well *coordinated*;

6) one *publishes* the measurements (after analysis).

Let us look at each of these points in some detail.

There is first of all the need for a good scientific motivation for undertaking the observational experiment, so one must find a phenomenon for which observation is scientifically needed, and which can be performed with the available instrument. Although at first glance this may be a difficult obstacle, in reality it is not. Our surrounding universe offers us such a variety of interesting and observable events, that there should be no one who cannot find a suitable project. There are many more stars that there are astronomers, and even by increasing the manpower in astronomy by several powers of ten (which is very unlikely to happen in the near future), there remains a very strong need for performing celestial observations.

Second, one cannot stress enough the importance of measurement in science. Observations in general serve to support, accept or reject physical theory. The early historical considerations about the structure of the universe were based on physical intuition derived from direct contact with our real physical world. Limited direct experience yielded an imcomplete view of reality, and only with the systematic application of increasingly powerful methods of measurement, did theory become more and more realistic. For example, pure human experience allows us to imagine sizes of natural objects (*e.g.*, celestial bodies) from 1 mm to about 10 - 100 km, whereas real dimensions estimated from measurement range from 10^{-30} m to 10^6 kpc. The first scale amounts to a ratio of 10^7 between smallest and largest size; the second scale covers a ratio of 10^{56}. A similar dramatic increase applies to the experience of mass, time, and maximum velocity.

Performing measurements is not a sufficient condition: it is of paramount importance that measurements are done with high precision. We all know that observations made by different observers at different instruments (or even at the same instrument at different occasions) often disagree. Furthermore, observations frequently happen at the limit of the instrumental capability, and the observational error, if quoted, is often grossly underestimated. Finally, instrumental time is *considered* to be too expensive for reconfirming a result by repeating the measurement.

The consequences of delivering poor and inaccurate data are disastrous, because there is a widespread opinion among theorists and observers that theory is merely a more or less disguised or sophisticated speculation, and that observations are solid facts. But due to this false conviction, errors in observations persist much longer than errors in theoretical work. One should not forget that theories often have sound and simple physical bases, and that it is easier to trace a wrong theoretical deduction than to prove that an observation is false.

Next to precision of observations, there is the important point of the frequency with which the data are collected. The literature too often shows that maximal information is extracted from too few data; in such a way, data are very often overinterpreted. It is absolutely necessary to take long strings of data, and to measure whenever possible (not just at selected phases). Also, one should always observe complete phenomena instead of partial events (*e.g.*, one should not only observe an eclipsing binary around the time of minimum, but cover the full eclipse, including the shoulders of the eclipse light curves).

One observer as an individual cannot always increase his observational potential to the level he eventually wishes. Therefore it is very important to coordinate efforts and eventually carry out observations in a small, well-organized team.

The last point, the dissemination of the data in the form of publication, is not only an absolute necessity, but also a reward. However, one should in principle only publish the data after a more or less elaborate analysis of the results has been carried out. It is only after such an analysis that one knows that the data are really valuable and useful. Unfortunately it becomes more and more difficult to get data papers published in regular journals.

The importance of these six considerations can be illustrated historically by taking the work of Tycho Brahe as an example. There was a strong scientific need for carrying out the astrometric observations for which he became famous. It is due to the aspect of measurement with new instrumentation that these observations made sense. The very high accuracy of his results has often been praised, but perhaps as important was the fact that he collected the data at any possible occasion and over a long interval. Finally the results were recorded and made available, and were used by Kepler to derive his laws.

According to Stebbins (1981), professional astronomers see amateurs as a group composed of a small core of serious individuals, surrounded by others who are essentially hobbyists. Hobbyists tend to be mainly occupied by designing and constructing mechanical and electronic components of their instruments, but they rarely get involved in serious observational science.

But for what concerns observers, I am convinced that there is no need to differentiate between amateurs and professionals: from the moment the observer is seriously motivated and masters the needed skills, no fundamental differences should exist.

But there are fundamental differences in background and in situation, and those contrasts tend to complement each other's possibilities rather than to eliminate the possibility of any valuable joint research. The most important aspects of this difference can be concisely discussed.

The professional observer as a rule has to travel a long distance to the observing site, whereas the amateur observer lives closer to his observatory (sometimes it is in his backyard). The direct consequence is that the professional observer has to plan his travel sometimes several months in advance, and often an observing run has to be cancelled because of lack of funds.

Also, since many observatories take visiting astronomers (some observatories work in a purely visitor-operated mode), professional observers find themselves in continuous competition for obtaining observing time. Frequently observing time is allotted in the wrong period (for example, during moonlit nights when instead dark time is required) or too late or too early in the observing season. The amateur, having his own instrument at his disposition does not suffer from this drawback.

It is certainly true that the total collecting surface of available professional telescopes steadily increases, but one should not forget that more telescope surface does not necessarily mean more openings for astronomical work. One astronomer alone can occupy a telescope, and then it makes no difference if the surface is 0.1 m^2 or one hundred times more. Figure 1 is taken from a report from the European

Science Foundation (1978, *A Study of Manpower in Astronomy* in the countries represented in the European Science Foundation), and illustrates the increase in number of telescopes available to the European Community for the period up to 1981. According to this source the ratio between the total available telescope surface and the total number of astronomers for a country like France amounts to about 0.06 m^2, which corresponds to a telescope diameter less than 30 cm, a middle-size amateur telescope. From the point of view of available collecting surface, there is no major difference between a professional and an amateur astronomer.

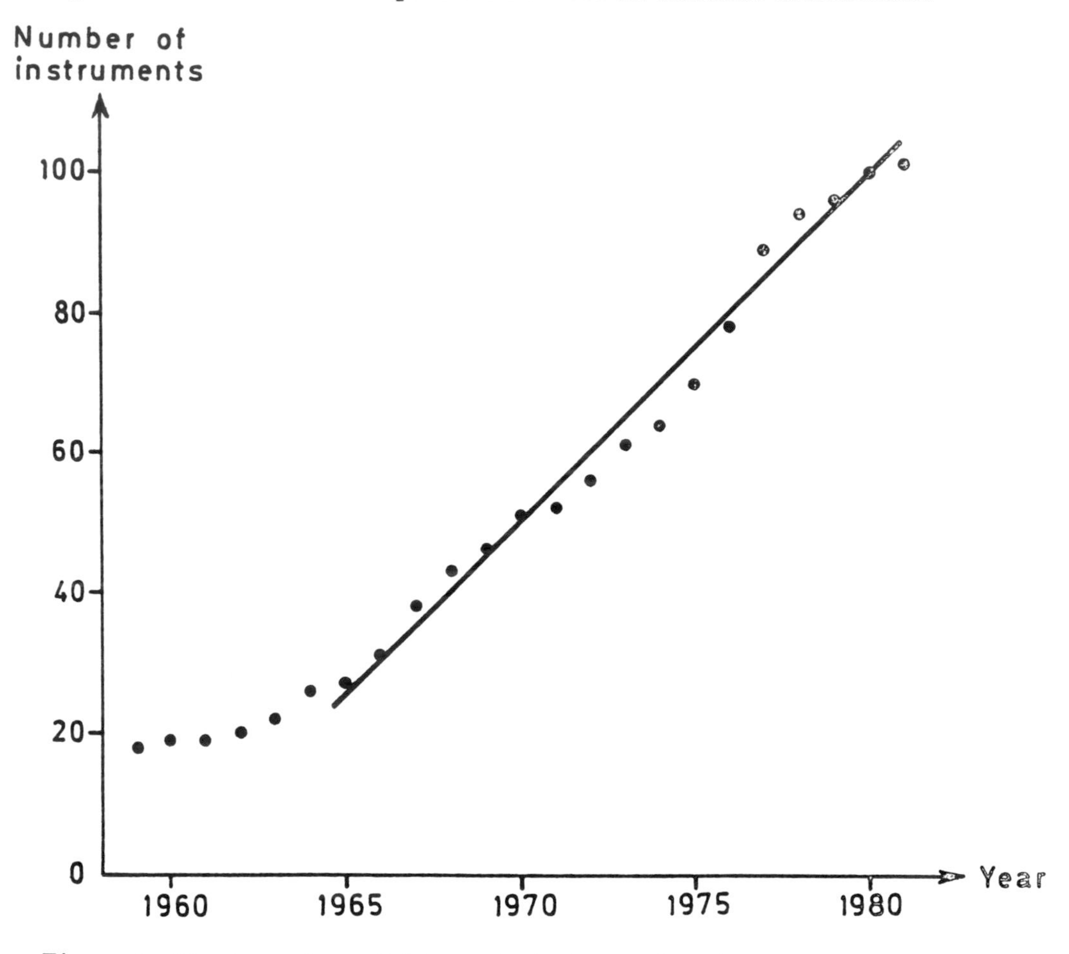

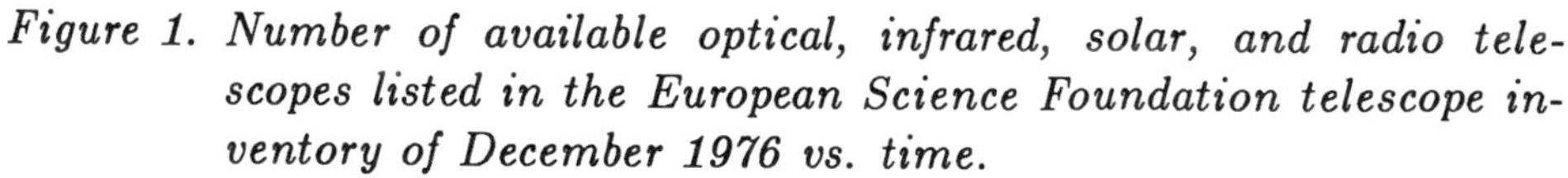
Figure 1. Number of available optical, infrared, solar, and radio telescopes listed in the European Science Foundation telescope inventory of December 1976 vs. time.

The fact that a professional observer acts as visiting astronomer has also other disadvantages. Whereas the amateur always works with the same instrument (which he knows very well), some professional astronomers continuously move from one instrument or instrumental configuration to another. In the best case they are really experienced observers who have only to compensate for some of the technical changes which were applied in between two successive visits (such as *e.g.*, filters, or photomultiplier tubes, or software support), but very often a less experienced observer (or even someone with no previous experience) gets at his disposal a complete optical, mechanical, and electronic configuration, finding himself in a dome with a number of black boxes that he marginally learns to use on a level of "which buttons to push to get something out". Such a situation of course is doomed to contribute to the production of poor data, and results in the further pollution of astronomical journals with doubtful results. This is one of the principal reasons why observers, even with the same instrument, obtain conflicting results.

There is also the conviction that good science can only be produced by using large instruments, and that the credibility of the work increases with the apparent magnitude of the observed objects. Therefore "small" (diameter 50 cm) professional telescopes are in less demand, and there is consequently a real danger that such undersubscribed instruments at a professional observatory may be put out of operation. (This has already happened at Kitt Peak National Observatory in the U.S.A.) On the other hand, there probably does not exist a single well-equipped professional telescope which is *small enough* to observe *e.g.*, Spica, one of the most interesting bright variable stars of the Southern Hemisphere. This star apparently has stopped its pulsation; therefore it should be monitored continuously in order to see if the pulsational behaviour has died out, or if it is only temporarily quiet. Unfortunately only a couple of professional observers look at it from time to time, using a telescope whose aperture is stopped down by a diaphragm or by neutral density filters. One needs no strong arguments to convince one's self that such cases are excellent targets for serious amateurs.

Finally there remains a very important difference between the situation of the amateur and the conditions of work of the majority of professionals. Other than the constraints which the amateur's personal professional activities put on his observational life, the amateur astronomer has an absolutely unlimited freedom of choice of the scientific work he wants to do. Once he has compiled a sample of scientific problems which he is able to tackle with the equipment he has, he is only limited by his personal choice, which in turn is determined by the attractiveness of the topic. The professional observer certainly has a comparable constraint from the side of his own profession (teaching, administration, search for funds, ...), but in addition to this he has only conditional freedom of choice of observing targets. This is especially the case for young astronomers, for whom the restrictions on the choice of the research topic come mainly from the principal research interests of the institute or team to which he belongs. Another very strong limitation is that some kinds of research allow one to write a paper more quickly and easily, whereas other research topics only allow one to write a short paper often after more than one or two years of work. Also, some types of observational work, (*e.g.*, astrometric work) are *considered* to be routinely executed observations with little scientific interest, and it is even difficult, if not impossible, to get these numerical results published in a major journal. This, together with the fact that such work is very tedious and time-consuming, keeps a lot of young astronomers from undertaking it, because they know very well that their job security is dominated by the principle of "publish or perish". This situation leads to an increasing tendency to carry out observing programs consisting of a number of very short observing sessions at a large telescope equipped with a sophisticated detecting device in such a way that each separate observing run leads to a new paper with superficial and hence speculative results. So the amateur, free from these constraints, could enter in fields which seem to be of minor interest to professional astronomers. This will not only result in the production of vital observations, but amateurs may also find out that such work gives them a lot of satisfaction. Astrometry for example combines celestial photography, the measurement of plates or films, and numerical calculation of positions.

All these differing aspects of the situation of both types of observing astronomers are strong arguments for intensive collaboration rather than being reasons for believing that there is little profit in joint work.

In what follows I briefly outline two of the projects in which I had the opportunity to work with several amateurs. Both projects concern the photometric observation of variable stars, but have a totally different approach: in the first one, amateurs contribute by observing a target during a worldwide campaign organized by professional astronomers; in the second case, observations are obtained by professional astronomers, and are analyzed by a team in which amateurs contribute.

I. COORDINATED MULTISITE OBSERVING PROGRAMS

Observing variable stars with short periods (a couple of hours to a couple of days) from sites located at different geographic longitudes is becoming more and more popular and is even regarded as a powerful method to collect data which can be obtained in no other way (see article by Stagg, later in this volume). One of the first international campaigns was the one conducted in 1956 by C. de Jager (de Jager 1962); the target was 12 Lacertae, a β Cephei star. The principal argument for undertaking multiple site observations of periodically variable stars is to eliminate the "alias" periods: the false periods which emerge from the data due to the fact that data strings are not long enough and as a rule contain very few contiguous cycles of variation. One of the aliases introduced, for example, is the one cycle per day alias, because the measurements are periodically interrupted by the sequence of day and night. Other aliases are introduced by the year (observing strings taken

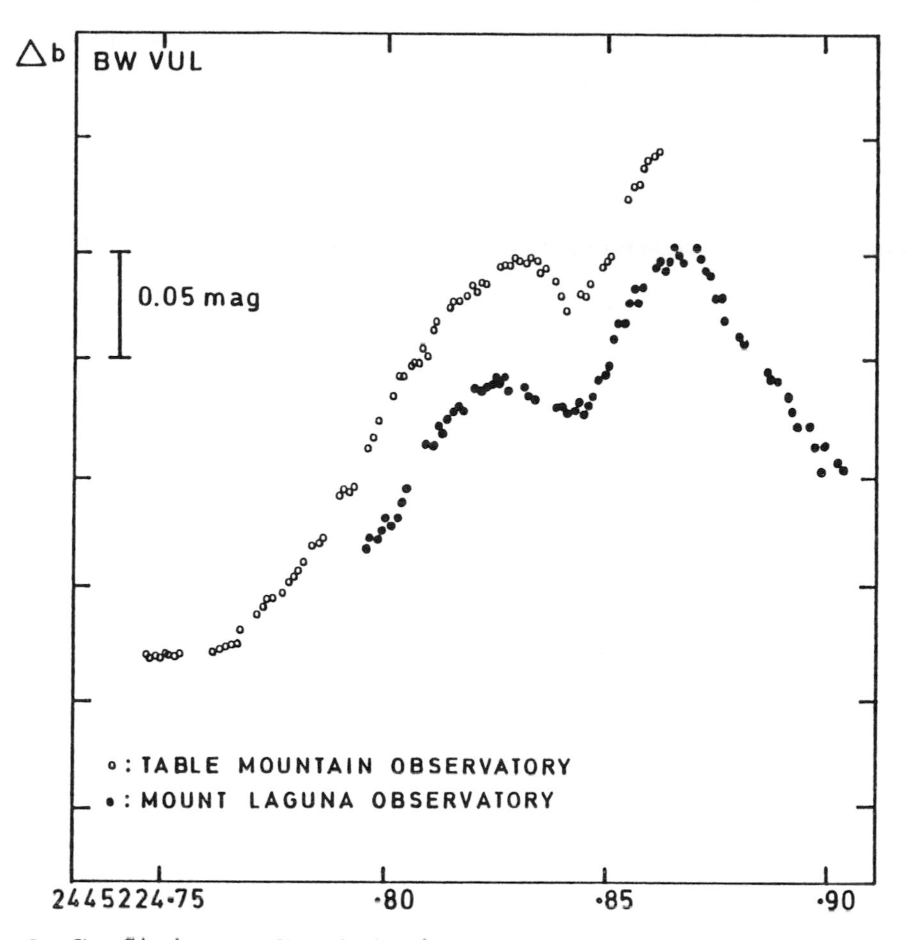

Figure 2. Conflicting results of simultaneous measurements at two different telescopes.

in successive years), by moonlight (cyclic variation of sky background), or even by the structure of the observing runs.

But the elimination of the aliases certainly is not the only strong argument for undertaking those campaigns. As an example the top curve of Figure 2 shows a portion of the light curve of the β Cephei star BW Vul, obtained on September 12, 1982, at Mount Laguna Observatory. The light curve shows the typical *stillstand* in light, a section of the rising branch of the light curve which displays more or less stationary light. But the lower curve, obtained simultaneously at Table Mountain Observatory, displays a much more indented stillstand phenomenon. All data were reduced by the same person using the same methods, though a difference, from unknown origin, is present. Only simultaneous measurements at different sites can give us such information; the frequency of occurrence of such divergences gives us an idea of the level of confidence of singly-observed events.

In the top curve of Figure 3, a light curve of BW Vul obtained on September 21, 1982 at Mount Laguna Observatory shows an interruption due to instrumental problems. Fortunately a light curve was observed during the same night at San Pedro Martir Observatory in Mexico. Due to the more eastern location of the latter, measurements were started about half an hour earlier than at the former, and they were accordingly terminated earlier. The San Pedro light curve not only gives extra data before the other observer could start, but the results also nicely fill in the unfortunate gap. Furthermore, the overlapping parts are of extreme value: they give the only objective estimate of the final precision of the data, and the more numerous the simultaneous data are, the better the statistical analysis of the precision that can be carried out.

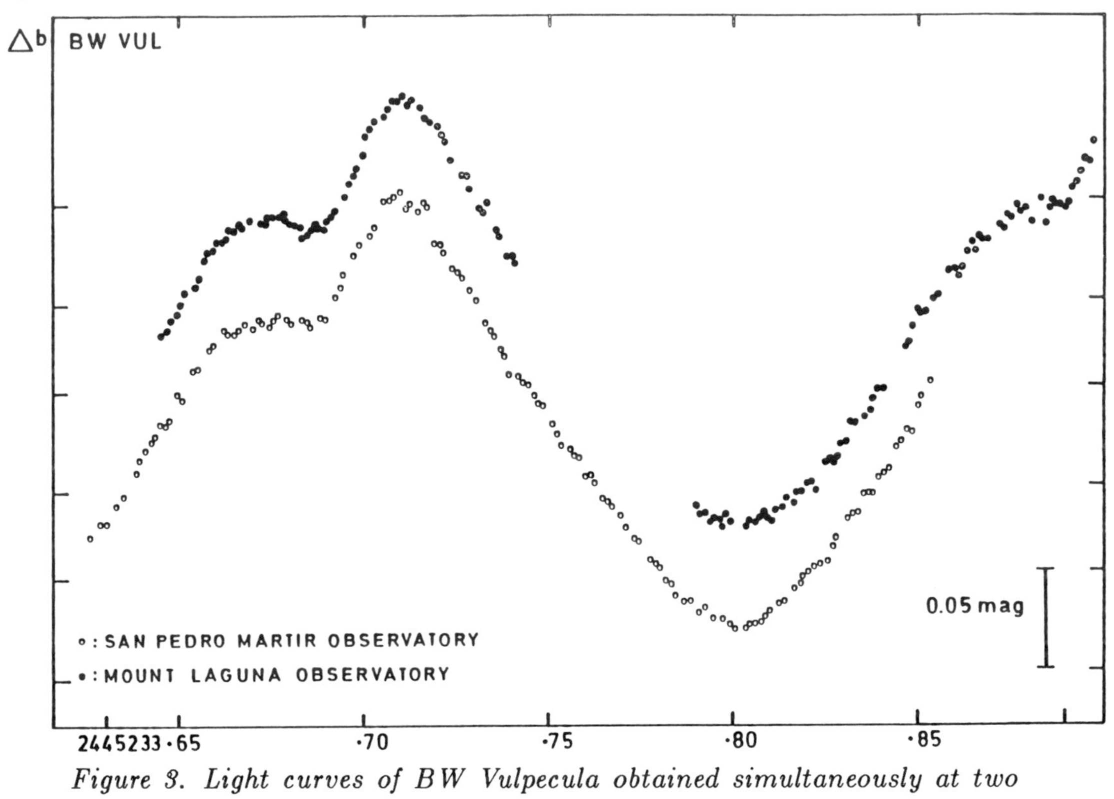

Figure 3. Light curves of BW Vulpecula obtained simultaneously at two sites.

Finally, Figure 4 shows a light curve resulting from the combination of results obtained at San Pedro Martir and at Mauna Kea Observatories. The larger the uninterrupted data string, the more the aliases in the period determination are suppressed.

Such campaigns of course have their specific organizational problems. The first one is the lack of good publicity: too few potentially interested people (am-

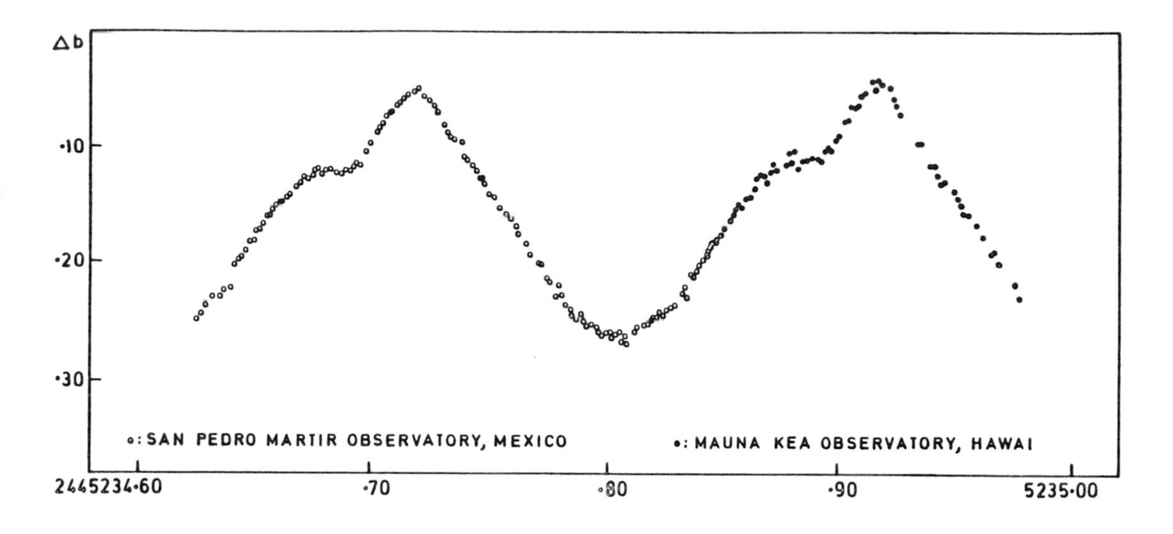

Figure 4. Extra-long light curve obtained from a combination of measurements obtained at two sites with different geographical longitudes.

ateurs or professionals) are aware that a specific campaign is planned. Another major problem is related to the data handling: central treatment of the data is an absolute requirement for obtaining homogeneous results, but unfortunately results are recorded in such a variety of ways and formats (magtape, papertape, listings, punched cards, floppy disks and cassettes of different sizes, ...) that this aspect represents a real nightmare for those who carry out the reductions. Unfortunately too few observers still participate in such campaigns, but there is hope that the situation will improve. As an extra argument for support of such work, I like to refer to the following statement.

"... I do not want to deny the importance of statistical relations that might be revealed by numerous but limited observations of a great number of objects. But it seems to me that real progress in the physical interpretation of a given type of variable depends essentially on detailed and at the same time complete and continuous observations of one typical star" (P. Ledoux, 1956 letter to C. de Jager).

II. LONG-TERM MONITORING OF VARIABLES

Many observers of variable stars are concerned with the study of stellar variability on long time scales (*e.g.*, the variations of supergiants, pre-main-sequence stars, long-period eclipsing binaries, etc.). In the past, the long-term observation of variables was very common, especially at observatories where whole teams of astronomers were collaborating on variable stars of similar type. But nowadays, where the bulk of observational work is done at visitor-operated observatories, real "monitoring" of variables during a time span of several years is almost impossible, mainly because allotted observing runs are too short and not contiguous. Principally because of changes in the instrumental system (*e.g.*, the change of a photomultiplier or of a filter set), long gaps between the visits of an individual observer inevitably affect the homogeneity of the results and contribute to deterioration of the precision. Therefore variable star observing programs with a long-term character are not often undertaken.

In 1982 a dozen active variable star observers belonging to the majority of the European member-countries of the European Southern Observatory decided to join efforts, and they started an international long-term collaboration with the aim of observing and studying a number of selected variables of different types over a time span of several years or perhaps of several decades. Such an observing program obviously calls for international collaboration in a well-organized team.

A number of individual observing programs were merged, and the most interesting stars were grouped according to several separate research topics. For each research topic, a principal investigator and a co-investigator were appointed. These persons are responsible for the coordination of the work within their own section. Table 1 lists all sections, and the corresponding investigators.

Section 7 consists of all objects which need immediate monitoring due to the occurrence of an unexpected event (flares or bursts) or due to exceptional observational possibilities (*e.g.*, simultaneous ground-based and space observations).

The principal investigator, together with the participants who are interested in obtaining measurements of a particular star, select the objects which are suitable for observation. The principal coordinator (C. Sterken) finally submits the application for observing time to ESO (this happens once every six months).

The actual list of objects contains about 150 stars, well distributed in right ascension. About 50% of them must be observed at a frequency of one measurement per night (throughout the corresponding observing season). The frequency for observing the remaining stars ranges between one measurement every second day to one measurement per month. The majority of objects have negative declinations, but some northern stars (maximal declination about 15 degrees) are also listed.

Observing our program stars makes sense only if measurements can be made frequently, and over long time intervals, so it is clear that only for the smaller telescopes one may hope to obtain sufficient observing time. Also, the participants have expressed a strong interest in using the Strömgren *uvby* photometric system

Table 1. Different sections of the "Long-Term Photometry of Variables". Principal investigators and co-investigators are listed.

Section	Principal Investigator	Co-investigator
1. Pre-main sequence stars	P.S. Thé Astronomical Institute University of Amsterdam Roetersstraat 15 1018 WB Amsterdam The Netherlands	H. Tjin a Djie Astronomical Institute University of Amsterdam Roetersstraat 15 1018 WB Amsterdam The Netherlands
2. Ap stars	H. Hensberge Astrophysical Institute University of Brussels Pleinlaan 2 1050 Brussels Belgium	J. Manfroid Institut d'Astrophysique Université de Liège Avenue de Cointe 5 4200 Cointe Ougrée Belgium
3. Eclipsing binaries	H.W. Duerbeck Observatorium Hoher List Universität Bonn 5568 Daun/Eifel F.R.G.	A. Bruch Astronomisches Institut Universität Münster Domagkstrasse 75 4400 Münster F.R.G.
4. Be stars	D. Baade European Southern Observatory K. Schwarzschildstrasse 2 8046 Garching bei München F.R.G.	
5. Supergiants	B. Wolf Landessternwarte Köningstuhl 6900 Heidelberg 1 F.R.G.	M. de Groot Armagh Observatory Armagh BT61 9DG Northern Ireland
6. X-ray sources	M. Burger Astrophysical Institute University of Brussels Pleinlaan 2 1050 Brussels Belgium	
7. Events of opportunity	C. Sterken Astrophysical Institute University of Brussels Pleinlaan 2 1050 Brussels Belgium	
8. Peculiar Late-type stars	F. Querci Observatoire de Toulouse 14, Avenue Edouard Belin 31400 Toulouse France	C. Zwaan Sterrenwacht Sonneborg Zonneburg 2 3512 NL Utrecht The Netherlands

(the *uvby* system is very well suited for the physical interpretation of the results; the intermediate bandwidths make transformations to the standard system easier and less subject to errors). The combination of an intermediate bandwidth filter system and a telescope of modest aperture (50 or 60 cm diameter) of course puts some constraints on the limiting magnitude of the selected objects (the faintest star is 11th magnitude), but the advantages of using such a system instead of *e.g.*, the *UBV* system are of higher importance.

Starting in October 1982, 20 observing runs have already been granted by ESO. Each run has a typical length of about three to four weeks. The observers are participants who volunteer to carry out the measurements according to the adopted observing scheme. The European Southern Observatory pays all costs related to the travel to Chile, the stay on the mountain, and the use of the instruments (the measurements are collected with either the Danish 50 cm telescope, the ESO 50 cm telescope, or the Bochum 61 cm telescope at La Silla). The administrative costs for the functioning of the project are paid by the Belgian Ministry of Education.

All measurements are obtained in a differential way using two comparison stars for every program star. Standard stars are also observed every night, so that all measurements can be transformed to the standard *uvby* system (for a description of the observing procedure, see Manfroid & Heck 1983). In order to preserve the homogeneity of the results, and also for practical reasons, all measurements are reduced centrally. Before leaving La Silla, the observer sends a magnetic tape with the recorded measurements to J. Manfroid at the University of Liège, Belgium. Experience has shown that the reduction procedure takes about two to four weeks, so that the principal investigators may receive the final results within one month after termination of the observing run.

This fast processing of the data has proven to be extremely useful, especially in those cases where unexpected changes of the light output of one of the stars calls for immediate action. Another advantage is that instrumental deficiencies are detected at once, and can be cured immediately.

The responsibility for the scientific value of the subprogram rests entirely with the principal investigator; he also redistributes the final data belonging to his section. All available data will be regularly published in the *Supplement Series* of an international journal. The submission of the data paper will only happen after the data are analyzed, the results are interpreted and the scientific discussion is published. The authors of the data paper are the individual observers who contributed to the measurements, and the scientific paper is published by those who applied for the observations and who carried out the research.

Since the program started, several scientific results are already available, and about 15 papers have been accepted for publication, or have been published.

The project offers several interesting possibilities. First of all, an important contribution to collecting valuable data is made. Second, the program induces a highly efficient and economical way of using telescopes of small to moderate size, and it also prevents inaction of useful instruments. For both professional and amateur astronomers it offers the possibility of interdisciplinary contacts. This is especially true for the youngest participants, who often have no clear research direction of their own; they especially profit from the guidance and background knowledge of the more experienced co-workers.

In principle any one can apply for data. Send your application with a clear description of why you wish data on a particular star, what precision you expect, and how long and how often we must observe, to the author or one of the principal investigators. Please remember that the applicant should select the comparison stars himself, and that he must prepare the finding charts.

My basic conclusion is that there are many fields in which amateur astronomers can contribute to scientific work, but that the principal cause for the lack of major collaboration between amateur and professional astronomers is caused by the poor exchange of information between both groups of people.

REFERENCES

de Jager, C. (1962). *Bull. Astr. Inst. Neth.*, **17**, 1.
Hall, D.S. & Genet, R.M. (1982). *Astronomy*, **10**, 24.
Manfroid, J. & Heck, A. (1983). *Astr. Ap.*, **120**, 302.
Mulholland, J.D. (1982). *Science News*, **121**, 2, 19.
Stebbins, R.A. (1981). *Journal of Leisure Research*, **13**, 289.
Stebbins, R.A. (1982) *Pacific Sociological Review*, **25**, 251.
Sterken, C., Vogt, N., Freeth, R., Kennedy, H.D., Marino, B.F., Page, A.A., & Walker, W.S.G. (1983). *Astr. Ap.*, **118**, 325.
Williams, T.R. (1983). *J. Amer. Assoc. Var. Star Obs.*, **12**, 1, 1.

THE INTERNATIONAL *UBV* PHOTOMETRIC CAMPAIGN ON Be STARS

John R. Percy
Department of Astronomy
University of Toronto
Toronto, Ontario
Canada M5S 1A1

The Be stars are hot stars (10,000 to 25,000 K) which have shown emission in at least one Balmer line on at least one occasion. The emission indicates that there is hot gas in a ring, disc, shell or cloud around the star. The Be stars are perhaps the most complex of all hot stars, and a satisfactory explanation for their emission and for their many kinds of variability has not yet been found.

Some years ago, Drs. Petr Harmanec, Jiri Horn and Pavel Koubsky (Harmanec et al, 1980a, b, c) proposed that a cooperative campaign be organized to gather long-term (10 to 20 years) *UBV* photometry of all Be stars with visual magnitudes brighter than about +6.5. (This would include over 200 stars; the Be stars make up a significant fraction of all bright stars.) The purpose of the campaign was to provide systematic information about the variability of a large, unbiased sample of objects, on time scales from hours to decades. Previous studies of Be stars had concentrated on a small number of objects whose behaviour was interesting but not necessarily representative. The list of campaign stars would deliberately include *all* kinds of Be stars, including supergiants and mass-transfer binaries. One of the benefits of the campaign would be that it would call attention to individual objects which needed special study. For example, about a quarter of the Be stars so far observed show variability on a time scale of a day or less, these stars need to be observed several times a night. There also appear to be some Be stars which can safely be observed less frequently than the others. A useful by-product of the campaign is a list of comparison stars of known constant magnitude, which can be used by future observers of Be stars — or of any other stars, for that matter.

The campaign initially included about 140 stars north of declination −20°, divided into approximately 70 groups. Each group contains one or more Be stars, a comparison and a check star similar in colour to the Be stars, and a "red standard". The purpose of the comparison star is to enable accurate *differential* photometry of the Be stars to be done. The check star and the red standard are also observed differentially against the comparison star, the first to check the constancy of the comparison star and the second to check the transformation of the differential magnitudes to the *UBV* system. Every effort was made to choose constant comparison, check and red standard stars, but inevitably a few of these have proven to be variable and have had to be replaced. This was expected: the organizers of the campaign were quite prepared to spend a year or more searching for optimal comparison stars for each group. Besides, the discovery of bright new variables is a worthwhile result in itself!

It was realized from the start that a project of this size could only be carried out in a cooperative way, with observations being made by many observers at many sites. This necessitated attracting potential observers, and this was helped by the fact that the campaign is sponsored by the International Astronomical Union's Working Group on Be stars. The Working Group also provided an effective channel for communication with participants, through its semi-annual *Newsletter* edited by Dr. Mercedes Jaschek of the Strasbourg Observatory.

The need to combine observations from many sources made it absolutely necessary to establish standard procedures for participation in the campaign: (i) observations must be made differentially using comparison, check and red standard stars which are obligatory for all participants; (ii) observations must be transformed rigorously to the *UBV* system using appropriate and well-documented procedures; (iii) comparison and check stars must be observed as frequently as the Be stars, in order to monitor the accuracy of the observations. In order to assist observers, and to maintain a uniformly high standard of data reduction, the campaign organizers have offered to reduce raw data using standard computer programs maintained at the campaign headquarters (Ondrejov Observatory in Czechoslovakia). The campaign organizers are also willing to work closely with individual observers, in order to better coordinate the observations. Several observers have taken advantage of these services.

Participants are free to publish their observations in the usual way, but they are strongly urged to send their manuscripts to the organizers *before publication*, to check that recommended procedures have been followed and that all necessary information has been included. They are also encouraged to send a copy of their observations to the campaign headquarters, where they are collected in a machine-readable archive, and can be made available to other observers for detailed joint studies of specific stars.

From the outset, it was realized that there were potential problems with the campaign. One was its sheer magnitude, and several astronomers initially recommended that the number of Be stars be drastically reduced, even at the expense of completeness. The campaign has been a great burden on the organizers, who have had to prepare the observing program and comparison stars, communicate with participants, maintain quality control, and supervise the archive. Another problem was the variability of some comparison stars but, as noted earlier, this was expected

— and produced useful benefits. Another problem is the limitations of the *UBV* system, especially for the study of hot stars. The *U* filter, for instance, is wide, and spans different parts of the spectrum of the star. This causes problems in the transformation and interpretation of the *U* magnitudes. Nevertheless, the *UBV* system is still the most widely-used photometric system; many previous studies of Be stars have been carried out in this system, and it was therefore decided to use it despite its limitations.

Some observers continue to observe Be stars outside the campaign, using different photometric systems, comparison stars and procedures. This is understandable, of course, but unfortunate, since their observations cannot be directly compared with those of other observers. It also results in some duplication of effort, especially if observers are not willing to share their data with others.

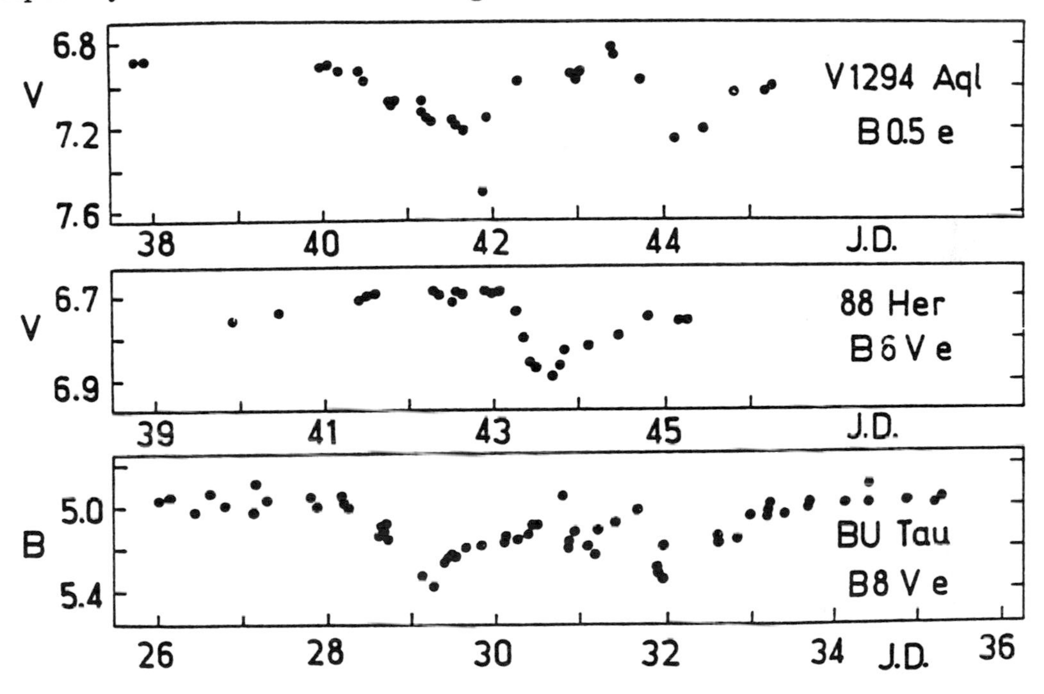

Figure 1. The long-term photometric variability of three Be Stars: V 1294 Aql, 88 Her, and BU Tau. Note that the time axis is marked in units of thousands of days, with JD 2400000 subtracted.

The campaign has now (1986) been operating since 1981, and considerable progress has been made. It has been extended to the southern hemisphere by the inclusion of about 80 more Be stars south of declination $-20°$, and a system of comparison stars has been established (Stagg, 1986). About thirty groups of observers initially expressed an interest in participating in the campaign, and about twenty groups in a dozen countries are presently active. Regular progress reports, including preliminary results, are published in the IAU *Be Star Newsletter*, and about a dozen research papers have been published which are directly or indirectly connected with the campaign. Several others are in various stages of completion. These will certainly help to encourage other observers to participate. The list of program stars is a useful resource in itself. So too is the list of comparison stars of known constant magnitude, which is of interest to both observers and to those with a special interest in standard stars.

The campaign has also served as a focus for Be star research in general. A long-term program of spectroscopic monitoring of Be stars is being carried out by Dr. Paul Barker in close cooperation with the photometric campaign. Other observers are using polarimetric, UV and IR techniques.

For all of these reasons, the campaign can be considered a success, and this success can be attributed to a number of factors: (i) the existence of a campaign centre and campaign organizers who have been willing to devote many patient years to the organization and supervision of the campaign; (ii) a rigorous set of procedures for making the observations and reducing them to the *UBV* system; (iii) the sponsorship of the IAU Working Group on Be Stars, and the availability of its *Newsletter*; and (iv) most of all, the scientific interest and merit of the campaign.

We encourage all interested and capable photometric observers to consider joining this campaign. We encourage also those who are studying Be stars using spectroscopic and other techniques to consider coordinating their observations with our photometric observations. Further information about the campaign can be obtained from Dr. Petr Harmanec, Astronomical Institute, Czechoslovak Academy of Sciences, 25165 Ondrejov, Czechoslovakia.

I am grateful to Drs. Petr Harmanec, Pavel Koubsky and Jiri Horn for their collaboration, and for their helpful comments on this article.

REFERENCES

Harmanec, P. (1980a). *Be Star Newsletter*, ed. M. Jaschek, Strasbourg Observatory. #1, 2.

Harmanec, P., Horn, J., Koubsky, P., Zdarsky, F., Kriz, S., & Pavlovski, K. (1980b). *Bull. Astron. Inst. Czechoslovakia,* **31**, 144.

Harmanec, P., Horn, J., & Koubsky, P. (1980c). *Be Star Newsletter*, ed. M. Jaschek, Strasbourg Observatory. #2, 3.

Stagg, C.R. (1986). *Ph.D. Thesis,* University of Toronto.

COMMENTS ON THE Be STAR MINI-CAMPAIGN OF OCTOBER 1983

Christopher Stagg
Department of Astronomy
David Dunlap Observatory
University of Toronto
Toronto, Ontario
Canada M5S 1A1

A single photometric telescope is ideally suited to studying two types of variable stars: the long period variables (periods much greater than a day) and the very short period variables (periods much less than a day). For the long period variables the procedure is simply to make a set of *UBV* measurements once every night or few nights until several complete cycles of variability are observed. For the very short period variables, on the other hand, one makes repeated observations to cover several cycles of variability in a single night. It is when one tries to observe variable stars whose periods are neither very long nor very short, that one runs into trouble. An example of this class are the "rapid" Be variables, whose periods generally range between about half a day and a day. The irregularity and small amplitude of their variations make them especially difficult to study. In a single night one cannot observe a complete cycle of variability, yet obtaining observations on several different nights does not solve the problem. A period analysis can be made, of course, but the periodograms contain many peaks corresponding to "pseudoaliases" of the period of variation. They differ from the true period by an integral number of cycles per day. Identification of the true period can be an impossible task. The only solution is to coordinate observations from stations at widely differing longitudes. In this way observations can be made almost around the clock, covering complete cycles of variability. The problem of pseudoaliasing then all but disappears. It was for this reason that Dr. J.R. Percy and the author initiated an international campaign on five northern Be stars for October 1983. This "mini-campaign" was carried out as part of the larger campaign described in the previous paper. Stations in Yugoslavia, the United States, Canada, and China were involved over a two week period, although earlier and later observations were also obtained. It was important to ensure that the same comparison and check stars were used by all groups, and that all observations were rigorously transformed to the *UBV* system. Without such precautions, the intercomparison of observations from different sources would have been difficult or impossible. With proper care, however, it was possible to combine the observations from all stations, and an analysis of the data is presently being prepared for publication (Stagg *et al.* 1986). Despite the very small amplitude of variability, at times only a few hundredths of a magnitude, periods and light curves were obtained for all five stars. The periodogram for the V magnitudes of o And, in particular, demonstrates how the problems of pseudoaliasing have been eliminated (Figure 1). This coordinated campaign by small photometric telescopes has thus been a great success, and a similar campaign has recently (late 1985) been carried out for Be stars in the southern hemisphere.

STARS OBSERVED

Program star	m_B	Preliminary period	Approx. V amplitude	Comparison star	Check star
o And	3.6	1.57 days	0.10 mag	10 Lac	HR 8733
KX And	7.3	0.47 days	0.10 mag	7 And	5 And
KY And	6.8	0.75 days	0.05 mag	HR 9011	HD 224166
LQ And	6.5	0.31 days	0.03 mag	7 And	5 And
EW Lac	5.1	0.72 days	0.20 mag	7 And	5 And

OBSERVERS

Station	Observers	Telescopes
Hvar, Yugoslavia	K. Pavlovski, H. Bozic	0.65 m
Kitt Peak, USA	J.R. Percy	0.4 m
McDonald, USA	L. Huang	0.91 m
Toronto, Canada	A. Fullerton, F. Schmidt	0.61 m, 0.48 m
Peking, China	W.S. Gao, Z.H. Guo	0.6 m

REDUCTION AND ANALYSIS

Toronto, Canada	C.R. Stagg
Ondrejov, Czechoslovakia	P. Harmanec, J. Horn, P. Koubsky, S. Stefl

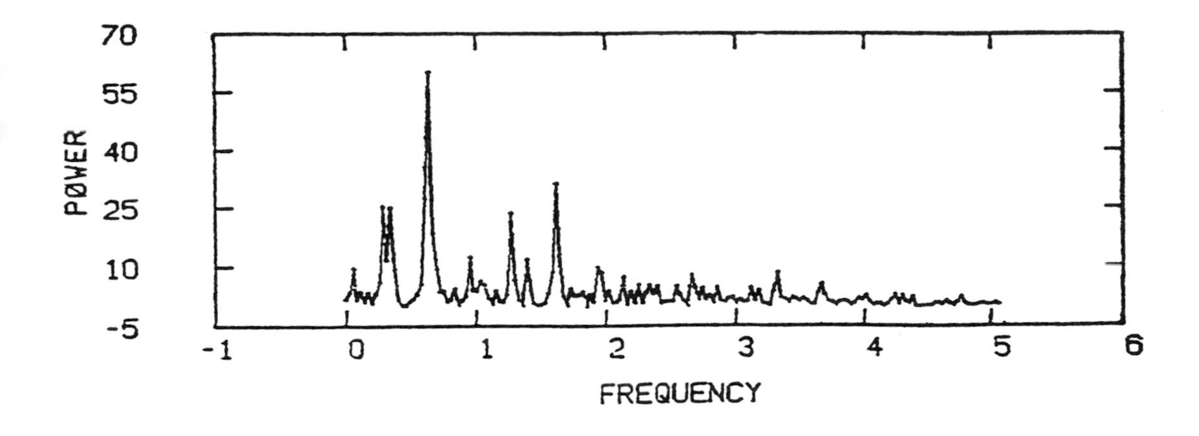

Figure 1. Preliminary periodogram for the V magnitudes of the Be star o And.

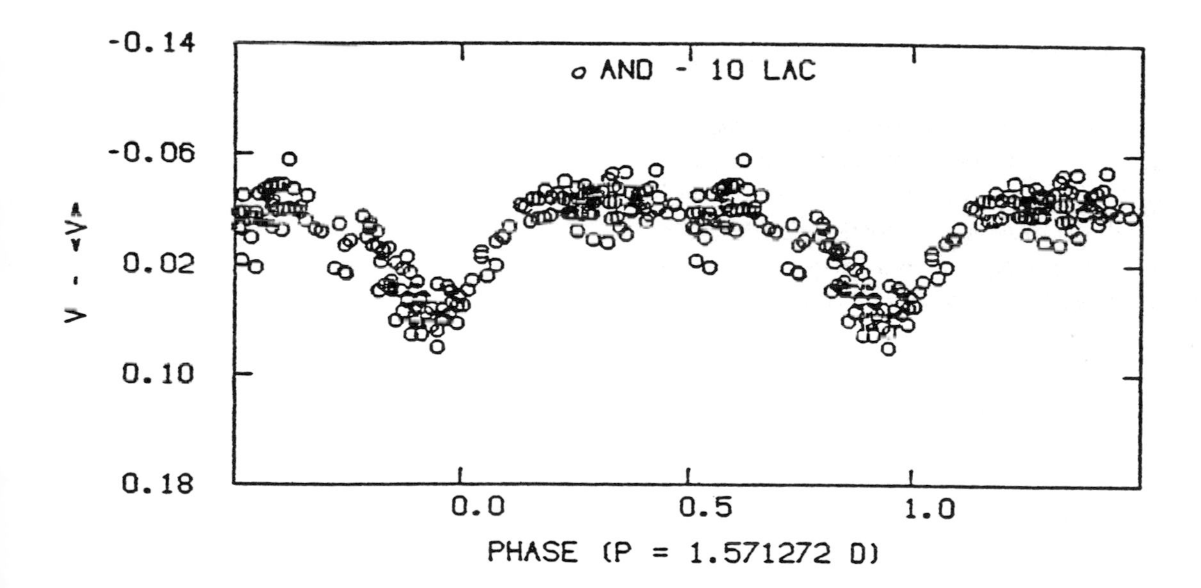

Figure 2. Preliminary V light curve obtained for the Be star o And. The large scatter is in part due to the problems of intercomparing observations made with different telescopes.

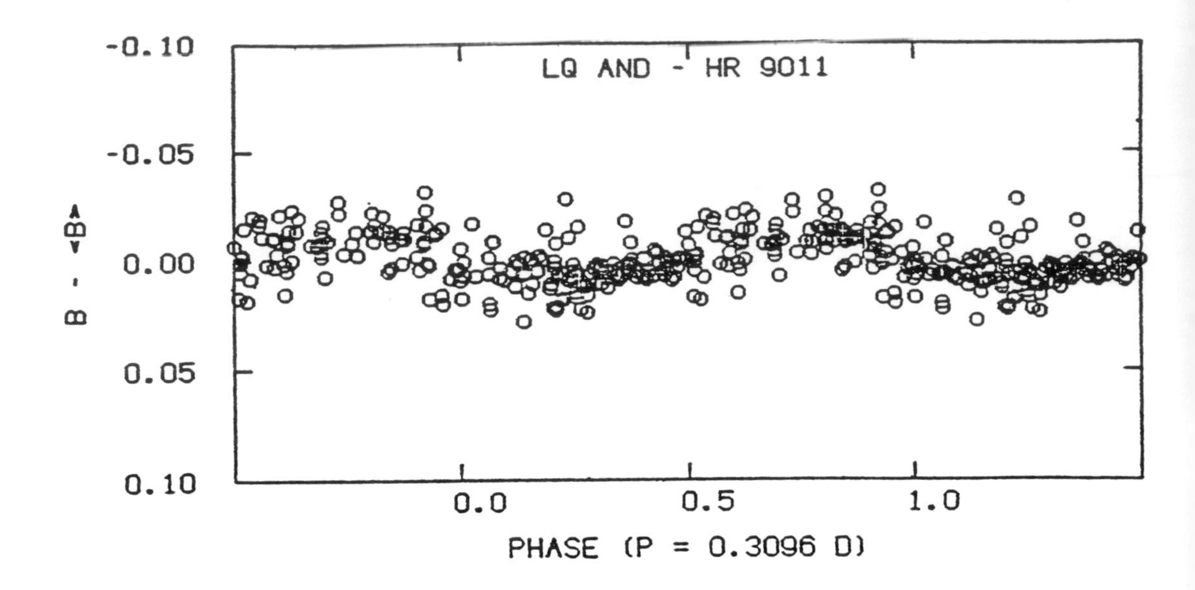

Figure 3. Preliminary B light curve obtained for the Be star LQ And.

REFERENCE

Stagg, C.R., Bozic, H., Fullerton, A., Gao, W.S., Guo, Z.H., Harmanec, P., Horn, J., Huang, L., Koubsky, P., Pavlovski, K., Percy, J.R., Schmidt, F., Stefl, S., & Ziznovsky, J. (1986). In preparation.

ARCHIVING OF VARIABLE STAR DATA — THE AAVSO EXPERIENCE

Elizabeth O. Waagen
The American Association of Variable Star Observers
25 Birch Street
Cambridge, MA 02138
U.S.A.

I. INTRODUCTION

The American Association of Variable Star Observers (AAVSO) was founded in 1911 as one answer to Argelander's plea for the systematic and useful monitoring of variable stars. Edward C. Pickering, a professional astronomer at Harvard College Observatory, was involved in the study of variable stars. He recognized the professional astronomer's problem of too many stars to observe and not enough time to observe them. William Tyler Olcott, a lawyer and avid amateur astronomer, wanted to do serious, valuable work with his telescope. The two collaborated to design the American Association of Variable Star Observers, creating an association which they hoped would be of significant help to professional astronomers and would be equally rewarding to the amateur astronomers involved. The past 74 years have seen their dreams bear fruit beyond their expectations.

With the establishment of an organization devoted to the gathering of data came the responsibilities which any would-be archiver faces and must manage: the data received must be recorded in an unambiguous format so that they may be properly interpreted at any later time; they must be stored in such a fashion that they may be accessed easily and conveniently at a later time; they must be handled in a consistent, methodical manner so that variances are not introduced into them and their value lost; some manner of dissemination of data must be established, whether actively, through publications, or passively, strictly on demand from researchers.

The AAVSO faced these responsibilities, and developed the procedures discussed below in answer to them.

II. ARCHIVING AAVSO VISUAL DATA — THE PAST

When the AAVSO was founded in 1911, a handful of intrepid and enthusiastic individuals began submitting data to Leon Campbell at Harvard College Observatory. Upon receipt, observations of each star were plotted on graph paper, adding to the running light curve of that star. The observations were then copied by hand onto ledger sheets — the permanent compiled archives for each star. Tabulations of observations were published each month in *Popular Astronomy* magazine. Each year, calculated dates of maxima and minima of long period variables were published in the *Harvard Circular*, and the next year's predicted dates of maxima and minima were published in the *Harvard Annals*.

All of the work involved in maintaining the AAVSO data archives was accomplished using the resources of the time — pencil and paper, adding machine, typewriter, and elbow grease. However, since the number of observations received was substantial but not overwhelming, the archiving process was reasonable and highly successful, as was the process of dissemination of information.

By the mid-1920's approximately 18,000 observations were being received each year — a substantial number, but still comfortably manageable using these procedures.

By 1935 the number of observations received had increased to the point where monthly publication was no longer practical. Instead, the approximately 50,000 observations being received were published quarterly in the *Harvard Annals*. However, the practices of hand-plotting and hand-ledgering continued unchanged.

Between 1935 and 1960 the number of observations submitted annually to the AAVSO fluctuated between 50,000 and 60,000. The archiving process continued unchanged, but some significant changes were made in the publication procedure. The AAVSO stopped publishing observations in *Harvard Annals* in 1946, and began publishing them every three months in *AAVSO Quarterly Reports*, following the format of the *Harvard Annals*. In 1950, the format of the *Quarterly Reports* was changed to better meet the needs of the astronomical community and the AAVSO. The *Reports* were published every 400 days, and included 10-day means for certain long period variables as well as listings of individual observations for other stars. Observations published in some of these *Reports* were also archived on microfilm.

With the explosion of interest in space and the space sciences in the early 1960's, the number of observations submitted to the AAVSO began to skyrocket. By 1965 over 80,000 observations were being received; by 1970, the total was up to 120,000 and had been as high as 140,000! While this tremendous growth in variable star observing was very exciting, it created serious problems for the AAVSO in managing and archiving the data pouring in. Financial constraints prevented the addition of staff members, and the process of ledgering by hand and plotting the observations directly from the reports became increasingly awkward.

In 1967 a new era began for the AAVSO with the inception of computerized data processing and archiving. Beginning with the data from 1961, observations were punched onto computer cards. The cards were mechanically sorted by star and date at Headquarters and read into a disk file on the Control Data Corporation computer at the Harvard-Smithsonian Center for Astrophysics in Cambridge, Massachusetts. A printout of the observations was obtained; this printout was used to plot the observations. The boxes of cards were stored at AAVSO Headquarters; no magnetic tape or permanent disk files were made for machine-readable access to the archives — the file was deleted after the printout was obtained.

In 1970, the *Quarterly Reports* were changed to include computer-drawn light curves showing 10-day means of long period variables. Data on other types of stars continued to be tabulated in the *Quarterly Reports.*

In 1976, the AAVSO acquired the funding and staff to implement the use of magnetic tape as a permanent storage medium for their data archives. The over 700 boxes of cards containing data from 1966 to 1976 were transcribed to magnetic tape at the Center for Astrophysics. The then-current data were sorted by star and time of observation before being taped, but the earlier data were not. From this time on, data were keypunched, sorted by star and time of observation, and transferred to magnetic tape on a monthly basis. The printout obtained each month continued to be used to plot the data by hand.

In 1978 the entire machine-readable archived database of the AAVSO had to be converted when the Center for Astrophysics changed their main computer system from that of Control Data Corporation to that of Digital Equipment Corporation. This project, along with the conversion of all of the AAVSO's software, took several months to complete.

In 1980 and 1981, the taped, unsorted data from 1966 to 1974 were ordered and combined with later data so that information on a particular star could be accessed more easily.

The AAVSO took another giant step forward in its ability to manage its ever-increasing database when, in December of 1981, in-house microcomputers were acquired. Software was developed, and observations were now entered onto 8 inch diskettes, rather than punched cards. This change made management of incoming data incomparably easier, and solved an increasingly serious problem of storage space.

In 1983, the method of preparation and format of the *AAVSO Reports* was changed. All editing of data and preparation of the light curves was done at AAVSO Headquarters, utilizing the AAVSO's microcomputers and software. The data were plotted with one point for each observation received, rather than each point representing a mean value for a number of days.

III. ARCHIVING AAVSO VISUAL DATA — THE PRESENT

The size of the monthly data base (at present, 15,000 to 20,000 visual observations are received at AAVSO Headquarters each month) exceeds the memory capacity of our computers, so we must rely on the computers of the Center for Astrophysics for large data processing projects. Also, the AAVSO uses magnetic tape as the permanent storage medium for its data, but there are no tape drives at Headquarters. Therefore, access to our machine-readable data archives requires the computers of the Center for Astrophysics.

Observers submit their visual observations on a monthly basis, using a standardized report form and reporting procedure. At the end of the month, incoming observations are sorted by observer and are checked by eye for apparent clerical errors. Utilizing customized software, observations are entered onto 8 inch diskettes using the Ithaca Intersystems microcomputers at Headquarters and are verified for entry errors. The CP/M-formatted diskettes are then converted to a DEC-compatible format and, using a special software utility, are read onto the hard disk of the DEC VAX 11/780. In a computerized two-step process, the observations are first checked against a list of AAVSO program stars for discrepancies, and are then sorted by star and Julian date. The computer-processed observations are examined by eye and any clerical errors (such as incorrect designations) are corrected so that future archival corrections are minimized. These observations are then copied onto magnetic tape. The unprocessed "raw" observations are also copied onto magnetic tape to insure future accessibility. Observations rejected by the computer programs are evaluated individually, and are inserted into the archives when indicated. Figure 1 is a graphic representation of the monthly data processing procedure.

Monthly files of observations are periodically merged together in order of star and Julian date, and these data for each star are added to our machine-readable archival files. Multiple copies of archival tapes are maintained to guard against loss of data through tape damage.

When data covering a specific time period are to be prepared for publication, pertinent observations are extracted from the archival files, copied onto diskettes, and brought to AAVSO headquarters, where they are converted to a CP/M-compatible format. Using custom, on-screen editing and graphics software, the observations are examined, and conservatively edited where necessary. Grossly discrepant points are removed, unless there are reasons not to do so. The observations are then plotted in a camera-ready format, read onto the disk of the VAX 11/780, and copied to magnetic tape for permanent storage. Figure 2 is a graphic representation of the procedure for preparing data for publications.

Current publication efforts take the form of *AAVSO Monographs*. Each monograph contains approximately twenty years of light curves on one star — all

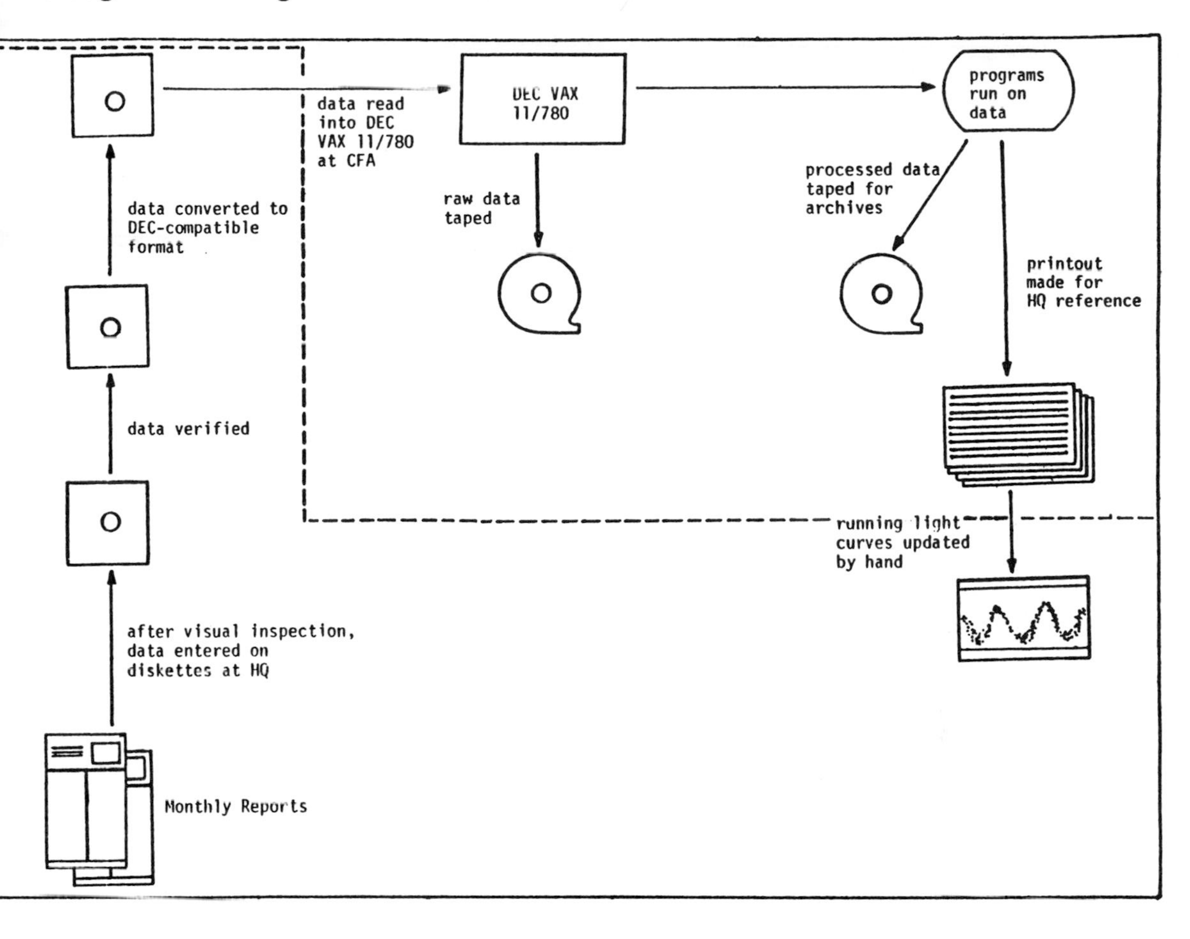

Figure 1. Monthly processing of incoming AAVSO visual observations. Steps above the dashed line occur at the Harvard-Smithsonian Center for Astrophysics; steps below the line occur at AAVSO Headquarters.

of the machine-readable AAVSO data up to 1985 are presented. Where possible, an earlier, historical light curve is included.

As of this writing, the total number of observations submitted to the AAVSO since its founding in 1911 is approaching 5.5 million. Of these 5.5 million observations, approximately 3.5 million, received since 1966, are in machine-readable form on magnetic tape. The other approximately 1.5 million observations are being entered on diskette for transfer to magnetic tape. This project is expected to take about 3 years to complete.

The AAVSO receives data from over 500 observers, about 350 of whom are very active. These observers come from all over the world — the international

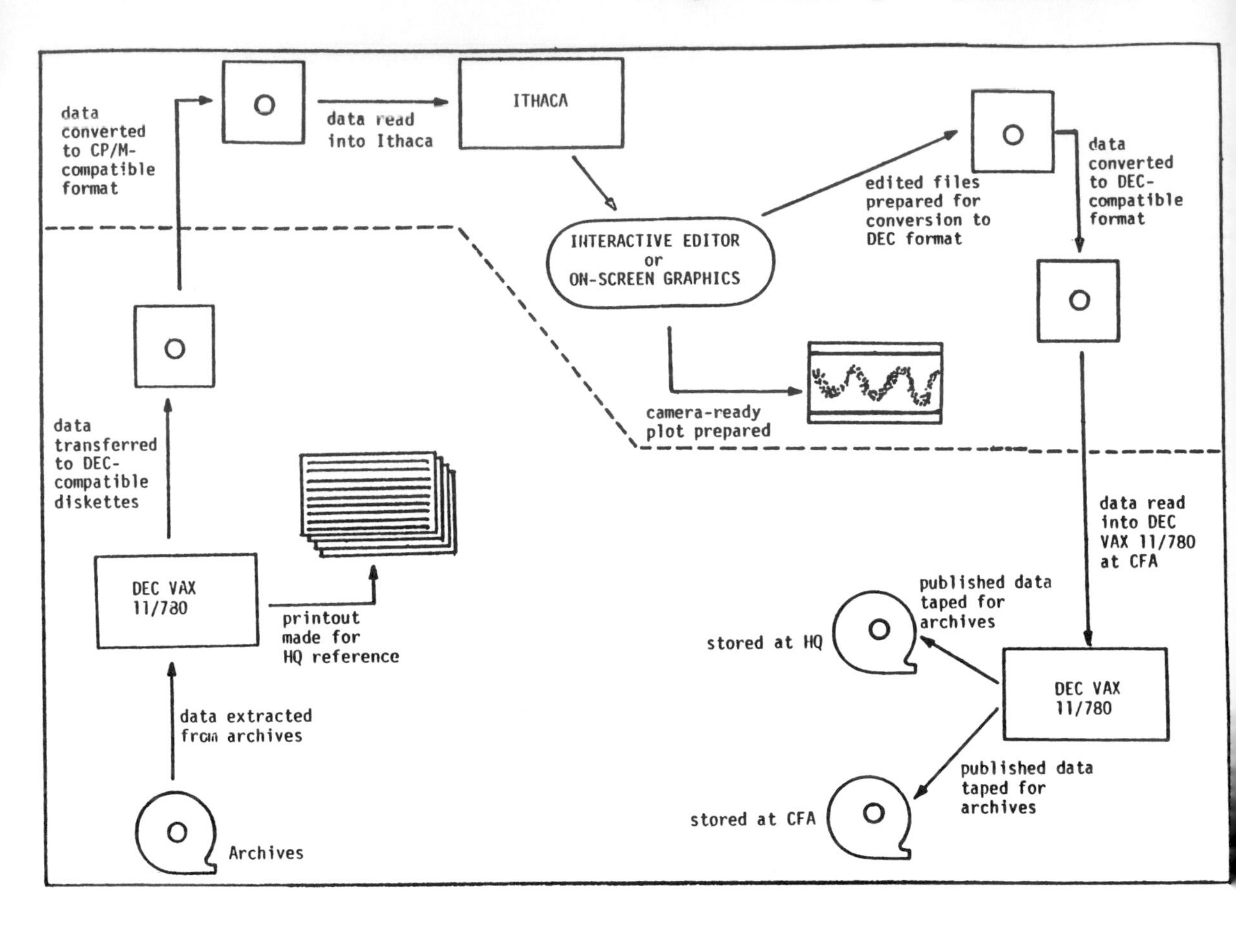

Figure 2. Preparation for publication of AAVSO visual variable star observations. Steps above the dashed line occur at AAVSO headquarters; steps below the line occur at the Harvard-Smithsonian Center for Astrophysics.

nature of the AAVSO is reflected in the fact that one-half of the data received today at the AAVSO comes from outside the USA!

IV. ARCHIVING AAVSO VISUAL DATA — THE FUTURE

It is expected that data will continue to be submitted largely on printed reports, through the mail. Some data will likely be sent via modem. It is not likely that much data will be sent via diskette because of the great variety of formats and operating systems present in microcomputers and the large amount of time required to convert many diskettes to a compatible format.

A modem will be added to our microcomputers to handle data sent by observers via computer. Other planned or projected additions to hardware include additional microcomputers, hard disk(s), a digitizer, an optical scanner, and, ultimately, a minicomputer with tape drives. This last piece of equipment would make us truly independent, allowing us to carry out all phases of archival management in-house. Software expansion and modification will occur to meet the evolving needs of AAVSO data archiving, retrieval, dissemination, and manipulation.

The machine-readable archival medium will continue to be magnetic tape. With suitable equipment, hard disk would make an admirable semi-permanent storage medium, giving opportunity for easy modification to the archives. A high-speed cassette backup for the hard disks would likely be employed. Another option is laser disk, an ideal medium for permanent electronic data storage. The problem with magnetic tape storage is that to assure the integrity of the data, tapes have to be copied periodically. With laser disk storage, there is no "expiration date" for the contents of the disk; it is truly a permanent medium. Unfortunately, it is not possible to update the contents of the disk without creating a new version of the disk altogether, an expensive process.

Data will continue to be disseminated via the different media available — publications of light curves and tabulations of data, magnetic tape written at the format and density required, diskette (8" CP/M or DEC and soon 5" CP/M formats), and printout. Expanded media for dissemination hopefully include direct transmission via modem, and, someday, perhaps, laser disk.

Incoming observations will continue to be made machine-readable within one to two months of their receipt at AAVSO Headquarters, and archived within several months.

Thus, with the completion of computerization of the archival data from 1911 to 1966, plus many observations of Argelander's and Argelander's students from the 1800's, the AAVSO will have 75 or more years of observations — 5.5 million of them — of variable stars in machine-readable form for future analysis.

V. ARCHIVING AAVSO PHOTOELECTRIC DATA

There are currently 60 stars in the AAVSO Photoelectric Observing Program. These stars have been chosen for any of several reasons, including small amplitude of variation (less than 1 magnitude), interesting features or behaviour, or at the request of an astronomer.

For each star, a special finder chart is prepared for use by all photoelectric observers. The chart indicates the variable, a check star, and a comparison star.

Also included on the chart are coordinates, magnitudes, and spectral information for the three stars. Photoelectric observers follow a standardized procedure — that is, they all observe the comp, check, and variable stars in the same sequence. This practice eliminates a possible source of error and therefore of scatter.

Unreduced observations are submitted on a monthly basis, using a standard form sent from Headquarters. Customized software, which keeps track of each observer's site and equipment, is used to reduce the raw data. Having all data reduced by the same program eliminates another major source of potential error, and allows data from different observers to be pooled in a more accurate analysis.

The reduced photoelectric data files are archived separately from the visual data files, as the two types of data are analyzed quite differently.

VI. ARCHIVING VARIABLE STAR DATA IN GENERAL

The many technological innovations of recent years are leading to exciting new opportunities for data gathering. Suddenly it has become possible to amass incredible amounts of data. Once the decision has been made to gather a certain type of data, great care must be taken in planning.

The technical considerations are clear: the apparatus to gather data must be designed; the procedure to gather data must be developed. This procedure should be systematic and standardized, in order to maximize the potential value of the data acquired.

Less straightforward to resolve is the "ethical" consideration — what is to be the fate of the acquired data? The data must be archived — it is a crime against science to gather unique data and not preserve them — and the data must be disseminated.

There are two options regarding disposition of data acquired: develop an archive and deposit the data in it, or send the data to an existing archive.

If the first option is chosen, the archive must be designed so that the data are available, easily accessible, and compatible with subsequent data. To this end, certain basic requirements must be fulfilled. The archive must: receive adequate, dependable, continued funding; have sufficient, able staff to manage the archive efficiently (if you don't have enough people, no matter how hard you try, it's extremely difficult to make any progress, and satisfy researchers' requests); have adequate physical facilities to keep the archives secure and viable; have adequate analytical equipment — computers — to keep the archived material manageable and accessible.

If the database grows beyond the limits of the archive, then either the archive must expand to accommodate the evolving database, or it must shut down. The contents of the archive should be deposited somewhere else, preferably in a long-established, active archive centre, and any subsequent data submitted to the defunct archive should be transferred as well, so that continuity is preserved. It doesn't matter as much *where* the data are archived (as long as the archivers are responsible) as *that* thcy are archived.

There are many kinds of variable star data being archived today — visual, photographic, photoelectric, CCD, etc. — and there will be more tomorrow. Not required is a single "mega-archive", where every observation of every variable star made using every technique is deposited, but it does make sense for one place to be the central archive for one type of variable star data. This arrangement has many advantages: the archivists can maintain a high, uniform quality across the entire database; redundancy through observations being deposited in more than one archive is avoided; potentially crucial data which might otherwise be overlooked is included, and so on. Certainly the researcher's task is simplified — he needs to go only to one place to obtain all the data of a certain type for a given star.

The fundamental ground rules described above apply to *any* archive, large or small, regardless of the type of data involved. However variable star data are archived, they must be scrupulously maintained for future generations of researchers.

Acknowledgements

I would like to express my sincere thanks to Dr. Janet A. Mattei, Director of the AAVSO, for many informative discussions and helpful comments. I would also like to thank Frances N. Withington for fruitful discussions.

REFERENCES

Campbell, L.C. (1926). *Variable Comments*, **1**, 60.
Campbell, L.C. (1931). *Variable Comments*, **2**, 45.
Hill, R.S. (1977). *J. Amer. Assoc. Var. Star Obs.*, **6**, 12.
Mattei, J.A. (1985). Private communication.
Mayall, M.W. (1972). *J. Amer. Assoc. Var. Star Obs.*, **1**, 47.
Olcott, W.T. (1936). *Variable Comments*, **3**, No. 9.
Percy, J.R. (1985). *AAVSO Photoelectric Photometry Newsletter*, **5**(3), 3.
Waagen, E.O. (1984). *J. Amer. Assoc. Var. Star Obs.*, **13**, 23.

VI. PERIOD ANALYSIS

The content of this section should be of interest to all readers of this book, if only because it deals with techniques which are essential to the correct interpretation of variable stars. The concepts involved are deceptively simple, even though in some methods the mathematics may appear difficult. In actual fact, other methods of period analysis use no more than high school algebra. The difficulty is in knowing what effect the observational error ("scatter") has on the results which the methods produce. In some cases, observational error can lead to periods which are incorrect or even non-existent.

Alex W. Fullerton's review provides a broad overview of the issues and approaches in period determination. For those of us who use period determination as part of our work, this review should be useful; for other readers, it should be interesting or even illuminating. The (O-C) method of period analysis is usually used to examine the correctness and constancy of a known period. Lee Anne Willson's review of this method is the most comprehensive published account known to me, and should be "required reading" for every student of variable stars.

Amateurs are normally more interested in making observations than in analyzing them, but this need not be the case. Many of them are more familiar with concepts of statistics than are professional astronomers (a sad commentary on contemporary science education), and many of them have access to computers. Perhaps these reviews will motivate a few amateurs to try their hand at period analysis.

Period analysis is a particularly suitable topic for students because it provides them with real data containing real errors, and introduces them to concepts such as periodicity, significance, curve fitting, and random and systematic error. These concepts are fundamental, and the mathematics is not difficult. It is fitting that Emilia P. Belserene has contributed a short paper to this section, because period analysis forms an important part of the training which the Maria Mitchell Observatory provides to its summer students. Like so many who have developed and implemented period analysis techniques, she has generously offered to make a copy of her program available to interested readers.

SEARCHING FOR PERIODICITY IN ASTRONOMICAL DATA

A. W. Fullerton
David Dunlap Observatory
Department of Astronomy
University of Toronto
Toronto, Ontario
Canada M5S 1A1

ABSTRACT

The goals and principles of astronomical period searching are reviewed. Particular emphasis is placed on the distinction between the detection of signals and the estimation of their properties. Several of the most commonly used algorithms are described in general terms, and their respective merits and deficiencies are discussed. Difficulties which hinder all period analyses are illustrated and some remedies are prescribed.

I. INTRODUCTION

The tricks and techniques employed to extract "hidden" periodicities from a series of observations rank among the most important tools used by observers and analyzers of variable stars. The importance of these calculations arises from the bias towards periodic phenomena shared by both observers and theoreticians, a bias so strong that "variable" is often used to mean "periodically variable". The observer appreciates periodic signals because they lessen his work load: having established the properties of a periodic variation, the star doesn't have to be observed intensely because its behaviour is (more or less!) prescribed. For the theoretician, both extrinsic and intrinsic periodic phenomena betray the existence of an underlying regularity which may be used to interpret the star's behaviour or to probe its structure.

This paper reviews the most popular methods of period searching in use today. The proliferation of digital computers has been a tremendous boon to the development of these algorithms and has greatly facilitated the computations they entail. Unfortunately, computers are not a panacea! The results of all period search programs are subject to a variety of difficulties arising from the inherent imprecision of the observations, the characteristics of the sampling of the data, and the "noise" generated by the algorithm itself. These difficulties and ambiguities are discussed in the last two sections of the paper. The statistical properties of the period finding programs offer the only protection against the detection of spurious periods.

Consequently, the need for rigorous statistical analysis is stressed throughout this paper.

II. PRELIMINARY CONSIDERATIONS

II.1 Input Data

The input data consists of a series of N_o observations of a dependent variable, $f(t_i)$, obtained at times t_i where i ranges from 1 to N_o. The times t_i are not necessarily equally spaced. In astronomical applications the dependent variable $f(t)$ is often the differential magnitude of an object in a particular filter or the radial velocity of an object. Each observation is contaminated by some random uncertainty ("noise") due to measuring errors, atmospheric fluctuations, guiding errors, and so on. This random variable is usually assumed to be from a Gaussian distribution characterized by a mean value of 0 and a standard deviation of σ_o. Noise processes are also assumed to be independent of frequency or "white", although this model is not always the most appropriate one.

II.2 Goals

There are generally two distinct goals associated with period analysis. In the first instance the input data must be searched for periods, and the statistical significance of each of the possibilities encountered must be rigorously tested. This is the problem of *detection*, and may be characterized by the question "Is there a statistically significant periodic signal in this data set?". In this context, periodic signals with a low probability (usually 1 per cent or less) of arising by chance are taken to be significant. The statistical properties of any period search method, especially those involved in the assessment of significance, are of crucial importance to the detection problem. For, as Morbey (1978) states: "only very rarely will there be an appropriately scaled assemblage of data where no periodicity at all is indicated. Even in apparently random data some periods will be found."

Presuming that a statistically significant signal is detected in a data set, the second goal of the analysis is to estimate the properties of the signal (e.g. period, amplitude) and to attach uncertainties to these estimates. This second goal is one of *estimation* and is typified by the question "What are the parameters associated with the signal which has been detected?". The determination of the best estimates of the signal parameters and their uncertainties will also involve the statistical properties of the period search method employed.

These two steps are not always separated by researchers undertaking period analysis. The distinction between "detection" and "estimation" is slight, but the questions which characterize each of the goals are quite distinct. More importantly, the statistics and frequency space sampling which are appropriate to "detection" are quite different from those appropriate to "estimation" (see II.3 below). Of course, not every period search requires the application of detection statistics. The existence of a periodicity may be known *a priori* from either independent observations or theoretical considerations. For example, once a periodic spectroscopic variation has been detected in an object, *any* photometric variation associated with this period is *automatically* significant. Similarly, if a star is known to be a Cepheid variable, the existence of a periodicity does not have to be demonstrated; only an "estimation" period search is required to establish the parameters of the light curve. In general, however, the two separate goals must be distinguished and the appropriate statistical analysis performed in order to prevent the proliferation of spurious periods.

II.3 Period Searching in General

To locate "hidden" periodicities, the input data are manipulated and tested by a function which is "sensitive" to periods in a manner described in the next section. At trial frequency (or period) grid points, the test function is evaluated. It attains an extremum (either a maximum or a minimum depending on the nature of the function) when a sympathetic trial frequency is encountered. A plot of the period search function against trial frequency indicates periods which may be present in the input data. Such a plot is often termed a "periodogram", although in this paper these diagrams will be called "generalized periodograms" to distinguish them from the class of functions described in section III.3). Once a generalized periodogram has been obtained, further analysis is required to determine whether any of the extrema are significant (*i.e.*, whether any significant periods are detected) or to estimate the properties of a known signal. In principle these procedures will work only for strictly periodic signals (see following article by Willson). Consequently, any changes in the period which occur must do so over intervals which are very long compared with the duration of the observations. In practice, however, these techniques remain useful for signals with periods which vary slowly and slightly.

The characteristics of the frequency (period) grid used to evaluate the period search function depend on the purpose of the analysis. For period searches designed to detect unknown frequencies which may or may not be present, it is crucially important that the frequency grid points be independent, or at least "quasi-independent" of each other (Scargle 1982). This property is required to ensure the validity of the statistical test used to assess the significances of extrema. In this regard, a generalized periodogram calculation is nothing more than a numerical experiment. In all experiments, whether physical or numerical, factors are varied independently of each other to determine relationships and causes. If the manipulations are not arranged so that each trial is independent of the others, then no conclusions can be reached concerning the effects of the factors on the system. Similarly, if neighbouring periodogram frequency grid points are not independent of each other, then no conclusions can be reached about the effect of a particular trial frequency on the system defined by the input time series. However, independence of the grid points is not a serious consideration if the periodogram is designed to estimate properties of a known signal. In this case, "oversampling" in frequency space is equivalent to interpolation, which is certainly permitted when best estimates of parameters are sought (Black & Scargle 1982; Scargle 1982). Thus an arbitrarily fine frequency grid *is* permitted for estimation purposes, but is *not* permitted for detection purposes.

There are further considerations concerning the frequency information present in a particular data set, both of which are determined by the temporal sampling characteristics of the data. It is intuitively clear that the highest frequency (*i.e.*, shortest period) information contained in a data set is determined by the intervals between successive measurements (Kurtz 1983). In the case of equally spaced data, 3 points are necessary to define a sine curve, hence the minimum period (maximum frequency) which can be detected is $2\Delta t$. The corresponding frequency is known as the Nyquist frequency: $\nu_{Ny} = \nu_{Max} = (2\Delta t)^{-1}$. However, the Nyquist frequency is not well defined for input data obtained at unequally spaced time intervals, the usual case in astronomy (Scargle 1982). Theoretically such a data set contains information on periodicities down to $\Delta t = \min(t_i - t_{i-1})$. In practice, however, a pseudo-Nyquist frequency may be defined by averaging $\Delta t = (t_i - t_{i-1})$, where large, uncharacteristic temporal gaps are avoided. Alternatively, the harmonic mean of all Δt may be used. The result is that a useful pseudo-Nyquist frequency may be defined by $\nu_{Ny} = (2\langle\Delta t\rangle)^{-1}$.

The final consideration concerns the longest period (smallest frequency) information present in a particular data set. Clearly this time is given by the total time spanned by the data, T. The corresponding minimum frequency is given by $\nu_{min} = (T)^{-1}$. This is the minimum frequency grid increment that any "detection" periodogram could utilize, providing that all the grid points are quasi-independent of each other. In the Deeming (1975) and Scargle (1982) Fourier-type periodogram techniques the "window function" contains information about the degree of coupling between grid points. Quasi-independent grid points are typically found to be separated by ν_{min}.

III. AN OVERVIEW OF PERIOD SEARCH TECHNIQUES

Three broadly defined classes of techniques are used to search for periodicities and to estimate their properties. These are a) string length methods, b) phase binning methods, and c) Fourier transform (or equivalently least squares or periodogram) techniques. Several variations usually exist within each class but they all function in much the same way.

III.1 String Length Methods

String length methods seek to minimize the total length of the line segments (termed the "string") which join adjacent observations in phase space (in essence, to maximize the "smoothness" of the curve). For a trial period P the phases ϕ for the input data are calculated from

$$\phi = FRC[(t_i - t_o)/P]$$

where t_o is the adopted epoch and where FRC indicates that only the fractional part of the expression is kept. Next the observations are put in order within the phase space associated with the trial period, and the string length is calculated according to the formula given by Dworetsky (1983):

$$l(P) = \Sigma[(f_j - f_{j-1})^2 + (\phi_j - \phi_{j-1})^2]^{1/2} + [(f_1 - f_n)^2 + (\phi_1 - \phi_n + 1)^2]^{1/2}$$

Clearly, the string length is calculated by repeated application of the Pythagorean Theorem in phase space. The function $l(P)$ is calculated for the complete range of trial periods, and plotted against P. The smallest value of l occurs for the true period, when all the observations lie along a smooth curve. The underlying curve may be of arbitrary shape.

There are two variations on the basic method given by Dworetsky (1983). The Lafler & Kinman (1964) method was the first to utilize modern digital computers. However, it only calculates the one-dimensional string length corresponding to the vertical distance between adjacent data points in phase space. Neglect of the contribution of the horizontal component of the string length prevents biasing by those regions which happen to be more densely populated, but it also increases the noisy appearance of the generalized periodogram. Renson's (1978) formulation includes a weighting function which allows for the unequal distribution of the observations in phase space.

The basic advantage of these methods is their simplicity. The major drawback they suffer is that for each trial period the phased observations must be sorted and placed in increasing order by phase. Although efficient sorting algorithms exist this procedure remains computationally intensive, especially for large input data sets.

III.2 Phase Binning Methods

Techniques which "fold" the input data using a trial frequency or period and examine the contents of the bins in phase space are frequently used in astronomical studies. Stellingwerf's (1978) Phase Dispersion Minimization (PDM) technique is particularly popular. The methods described by Jurkevich (1971) and Morbey (1973, 1978) are conceptually similar to it.

In the PDM method, phase space (0.0,1.0) is divided into a number N_b of compartments ("bins"), typically 5, each of equal extent. The input data are phased with a trial frequency and are distributed among the bins accordingly. Usually a number, N_c, of overlapping bins are considered. The various sets of bins ("covers") are designed so that the outcome of the calculation will not depend on the distribution of the input data in the bins. A typical bin structure is characterized by $(N_b, N_c) = (5, 2)$; this structure is illustrated schematically in Figure 1. Each of the input data points falls into N_c bins.

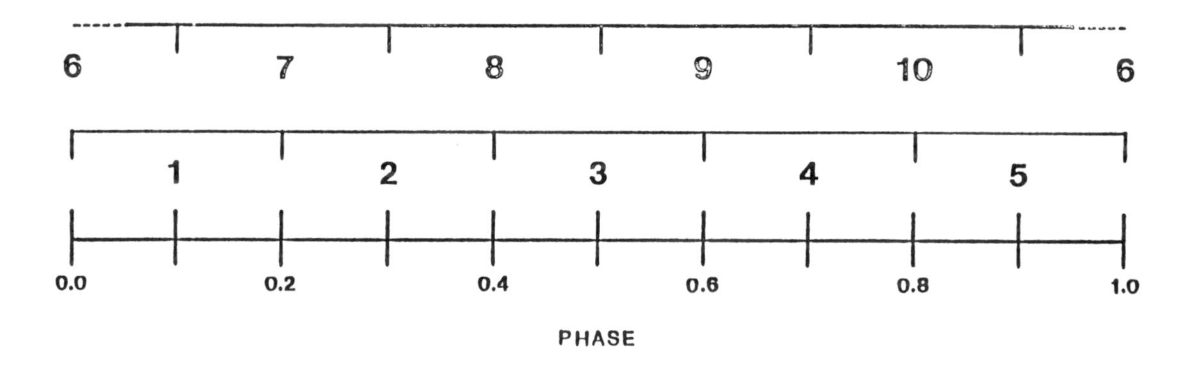

Figure 1. Schematic of PDM (5,2) bin structure. Phase space is divided into 5 equal bins, each extending for intervals of 0.2 in phase. There are 2 sets of "covers"; each set is offset from the other by 0.1 in phase. The 10 resulting bins are labelled. Clearly each data point falls into 2 bins.

For a particular trial frequency, the variances of the data assembled in each of the $N_b N_c$ bins are calculated and summed. The sum of the bin variances is compared with the global variance of the input data. Most trial frequencies will produce a random distribution of data points in phase space and so the ratio of the total bin variances to the global variance will be approximately unity. However, for the trial frequency which corresponds to a true period the scatter within the phase bins will be reduced dramatically and the ratio of the two variances will be a small number. In a plot of the ratio of the two variances *vs.* frequency the frequency of the true periodicity is indicated by a deep minimum. The depth is an indication of the significance of the fit for *it reveals the degree to which the scatter about the mean light curve is reduced.*

Like the string length methods, phase bin techniques make no assumptions about the shape of the variation being studied. However, the phase bin methods are not usually as computationally intensive as the string length methods, primarily because the data do not have to be ordered within the phase bins.

III.3 Fourier Transform, Periodogram, and Least Squares Methods

The Fourier transform and related techniques are all based on the discrete Fourier transform and employ the elegant concepts embodied in harmonic analysis. Although apparently distinct, all of these period search methods are conceptually similar. Fourier Transform and periodogram techniques are very closely related, for the latter are essentially the squared modulus of the former, suitably normalized (Deeming 1975; Scargle 1982). Least squares fitting techniques are formally equivalent to periodogram analysis under certain circumstances (Scargle 1982).

The current popularity of these techniques is largely due to the computational efficiency of the Fast Fourier Transform (FFT) routine. Unfortunately this algorithm requires input data which are equally spaced in time, an unusual circumstance in optical astronomy. Even in cases where the FFT is applicable (*e.g.*, Middleditch & Nelson 1973) the periodogram or "power spectrum" is usually calculated, at the expense of some phase information.

An extremely useful formulation of the periodogram function has been developed by Scargle (1982):

$$P_X(\omega) = \frac{1}{2}[C^2(\omega) + S^2(\omega)]$$

where

$$C(\omega) = A(\omega)\Sigma X(t_j)\cos(\omega[t_j - \tau])$$

$$S(\omega) = B(\omega)\Sigma X(t_j)\sin(\omega[t_j - \tau])$$

$$A(\omega) = (\Sigma\cos^2(\omega[t_j - \tau]))^{-1/2}$$

$$B(\omega) = (\Sigma\sin^2(\omega[t_j - \tau]))^{-1/2}$$

and where $\tau(\omega)$ is defined by

$$\tan(2\omega\tau) = (\Sigma \sin 2\omega t_j)/(\Sigma \cos 2\omega t_j)$$

In these expressions, ω is the angular frequency, $\omega = 2\pi\nu = 2\pi(P)^{-1}, X_j$ are the input data obtained at arbitrarily spaced times t_j, $j = 1$ to N_o, and $\tau(\omega)$ is a phase function designed to make this periodogram formally equivalent to least squares fitting techniques. To understand how this function is sensitive to frequencies present in the data, suppose that the data set X_j contains a sinusoidal signal of angular frequency ω_o. When the trial frequency is not close to ω_o the input data X_j and the $\cos(\omega[t_j - \tau])$ and $\sin(\omega[t_j - \tau])$ terms are out of phase. The terms in the summations have random signs with respect to each other and they largely cancel out. However, when $\omega \simeq \omega_o$ the data and the sinusoidal factors are nearly in phase and the terms in the summations add coherently so that P_X attains a relatively large value. Thus the periodogram reveals hidden periodicities by "resonating" with them. The Scargle formulation is somewhat more complicated than its predecessor, the famous Deeming algorithm (Deeming 1975), but it possesses very desirable statistical properties which make its use preferable (Horne & Baliunas 1985). An example of a periodogram obtained by this method is contained in the paper by Stagg, elsewhere in this volume.

By design, the calculation of a Scargle periodogram is equivalent to fitting sinusoids to the data by the method of linear least squares. Least squares techniques are conceptually straightforward; descriptions are given by Vanicek (1969, 1971), Lomb (1976), and Ponman (1981). At each of the trial frequencies ω a linear combination of sine and cosine functions with argument $(\omega t - j)$ are fit to the data by least squares. The sum of the squared amplitudes from both of these terms is plotted against frequency to form a graph which closely resembles a periodogram (*cf.* Lomb 1976, Figures 1 and 2). Peaks in the graph indicate periods; the heights of the peaks indicate significances through their relationship with the F statistic.

In addition to the linear least squares approach (where the frequency enters as a parameter), nonlinear least squares calculations have been used to solve iteratively for the period(s) and amplitude(s) which best fit the data (*e.g.*, Wizinowich & Percy 1979). These methods require good initial estimates for the period to prevent convergence to a spurious period located in a local rather than the global minimum in the parameter space.

Fourier techniques and their relatives differ from the other two classes of methods in that the signal is assumed to have a definite functional form (sinusoidal). This would appear to limit the usefulness of these techniques for the analysis of nonsinusoidal variations (*e.g.*, Cepheid light curves). In principle Fourier methods are applicable to arbitrary waveforms through the Fourier Decomposition Theorem which states that any periodic signal can be decomposed into a (possibly infinite) summation of sinusoids. The frequency of each of these sinusoids will be an integral multiple of the fundamental frequency ν_o, *e.g.*, $2\nu_o$, $3\nu_o$, $4\nu_o$, etc. For example, adequate representations of Cepheid light curves usually require summation over the first 4 to 8 harmonics (Moffett & Barnes 1985). Many more Fourier components are required to reconstruct the light curve of a typical eclipsing binary.

However, some practical difficulties arise owing to the effects of aliasing and noise (see section V) and there is considerable risk that the higher harmonics may not be recovered. Under these circumstances the other "non-parametric" methods provide better estimates of the properties of the signal. However, for detection purposes Fourier methods remain valid even for significantly nonsinusoidal signals (Black & Scargle 1982). Despite the distribution of the signal over more than one Fourier component, most of the "power" in curves typical of Keplerian orbits, for example, remains at the fundamental frequency (Jensen & Ulrych 1973). Thus the presence of at least the fundamental frequency should be detectable in any data set of adequate quality (with the exception of many eclipsing binary light curves) although other techniques may be required to estimate the harmonic content of the waveform.

IV. TESTS AND COMPARISONS

From the descriptions of their operation, it is clear that each of the three classes of period search techniques is able to estimate the properties of periodic signals. Series of tests performed by Perez de la Blanca & Garrido (1980) and by Heck, Manfroid, & Mersch (1985) on various subsets of techniques confirm this capability. The tests performed by Heck, Manfroid, & Mersch are particularly impressive. These researchers concluded that the non-parametric PDM and "Renson" methods are slightly (probably insignificantly) preferable to the other methods though they also comment that non-parametric and Fourier methods are complementary to some degree. Similar tests performed by various workers at the University of Toronto give comparable results, although a preference for periodogram techniques has been established. Heck, Manfroid, & Mersch find all the methods to be roughly equal in computational efficiency with the exception of the PDM method which is slower because of its complicated bin structure.

However, the equivalence among these techniques vanishes when the problem of detection is considered, primarily because of the lack of appropriate statistics for all methods except the Scargle periodogram (and, by extension, least squares fitting). The factor which the other methods fail to account for is the "statistical penalty" exacted for examining many frequencies (Scargle 1982). The origin of this penalty is best understood by considering a periodogram (now taken in the general sense) to be a numerical experiment. Since the "experiments" a periodogram performs are inherently noisy, large random fluctuations are expected to occur. Clearly if the experiment is repeated often enough (*i.e.*, if enough trial frequencies are searched) the chance of encountering a noise spike of any specified height increases and approaches certainty as the number of experiments tends to infinity. If one large periodogram peak is uncovered among the N frequencies examined, the fact that there were $(N-1)$ non-detections must be accounted for when the significance of the peak is assessed.

Within the context of Scargle's modified periodogram the statistical penalty for searching many frequencies is embodied in the threshold height which a peak must attain to be considered significant. This height is usually expressed in units of the noise, *i.e.*, it is a power signal-to-noise (S/N) ratio. As the number of frequencies searched increases so must the height of a peak in order to achieve a specified degree of statistical significance. Thus, if z is the power S/N threshold for 1 per cent significance, then z is given by

$$z(N; 0.01) = 4.6 + \ln(N)$$

If the detection periodogram is calculated at 100 frequency points the power S/N threshold for 1 per cent significance is 9.2 (corresponding to an amplitude S/N ratio of 3.0); if 500 frequencies are searched this number becomes 10.8 (amplitude S/N ratio of 3.3). Of course each of the frequency grid points must be quasi-independent in order for this test to be valid (section II.3).

Only the statistics of the Scargle periodogram account for the statistical penalty. Heck, Manfroid & Mersch (1985) have also recognized that the other methods are inappropriate for detection purposes, for they state that "these techniques should then stay in an essentially descriptive scheme. When ν has to be estimated none of these criteria has a known distribution, not even Fourier's". The last comment refers to the Deeming periodogram; Scargle's modifications remedy this difficulty.

The reasons for these shortcomings lie in the complex, essentially undetermined, statistical properties of the non-parametric methods. These properties are exceedingly difficult to evaluate analytically. Consider, for example, the PDM method. Individual data points occupy more than one bin, consequently the various bins are not independent of each other! Rigorous detection statistics are probably not derivable under these circumstances. The only approach available for the non-parametric functions is to undertake extensive computer simulations with random data (obtained by shuffling the temporal order of the actual data, for example) to derive "empirical" significance estimates (Nemec & Nemec 1985). An advantage of these simulations is that they make no assumption about the distribution of the noise. However, they can be quite costly in computer time, particularly when large data sets are involved.

V. COMPLICATIONS

Unfortunately, all the procedures for performing period analyses described in the preceeding sections suffer from several complications. These difficulties arise from the imprecision inherent in the input data, and from interactions with the sampling "window" through which the data are obtained. They may be overcome under many circumstances.

V.1 Aliasing and Pseudo-Aliasing

The temporal sampling characteristics of the input data give rise to insidious difficulties known as aliasing and pseudo-aliasing. As the term suggests, "aliasing" inhibits and sometimes precludes the identification of the correct signal frequency. Instead a related "alias" frequency is detected. This phenomenon arises when the input data are sampled at equally spaced time intervals, Δt. This temporal regularity permits sinusoids of frequencies $\nu_o \pm n(\Delta t)^{-1}$ to fit the data equally well (Figures 2a) and 2b)); in this expression ν_o is the true signal frequency and n is an integer. Fortunately, astronomical observations are seldomly spaced equally in time, usually because of the follies of the weather. Even small deviations from equal sampling are sufficient to make aliasing unimportant (Figure 3). As a result, aliasing is not usually a problem for astronomical time series.

However, a similar phenomenon termed "pseudo-aliasing" by Scargle (1982) (though most people do not distinguish the two phenomena) is almost always present in astronomical observations. Pseudo-aliasing causes the same sort of confusion over the identity of the true period as genuine aliasing although it originates with the cycle-count ambiguities which exist in gapped data. Put another way, pseudo-aliasing is caused by the presence of a regularity in the overall observing pattern. For example, the periodogram of observations obtained on consecutive nights will show strong ± 1 d^{-1} pseudo-aliases; observations made on an annual basis possess ± 1 yr^{-1} pseudo-aliases, and so on. Without resorting to gap-filling algorithms (see V.2), the only ways of avoiding pseudo-aliasing consist of restricting attention to observations of rapid phenomena on a single night or arranging for essentially continuous unequally spaced observations from a satellite, from telescopes situated at a variety of longitudes, or from a telescope located at one of the Earth's poles (during the appropriate season, of course!).

Although pseudo-aliasing can make period analysis impossible, there are usually ways of dealing with it. This is particularly true of Fourier techniques which

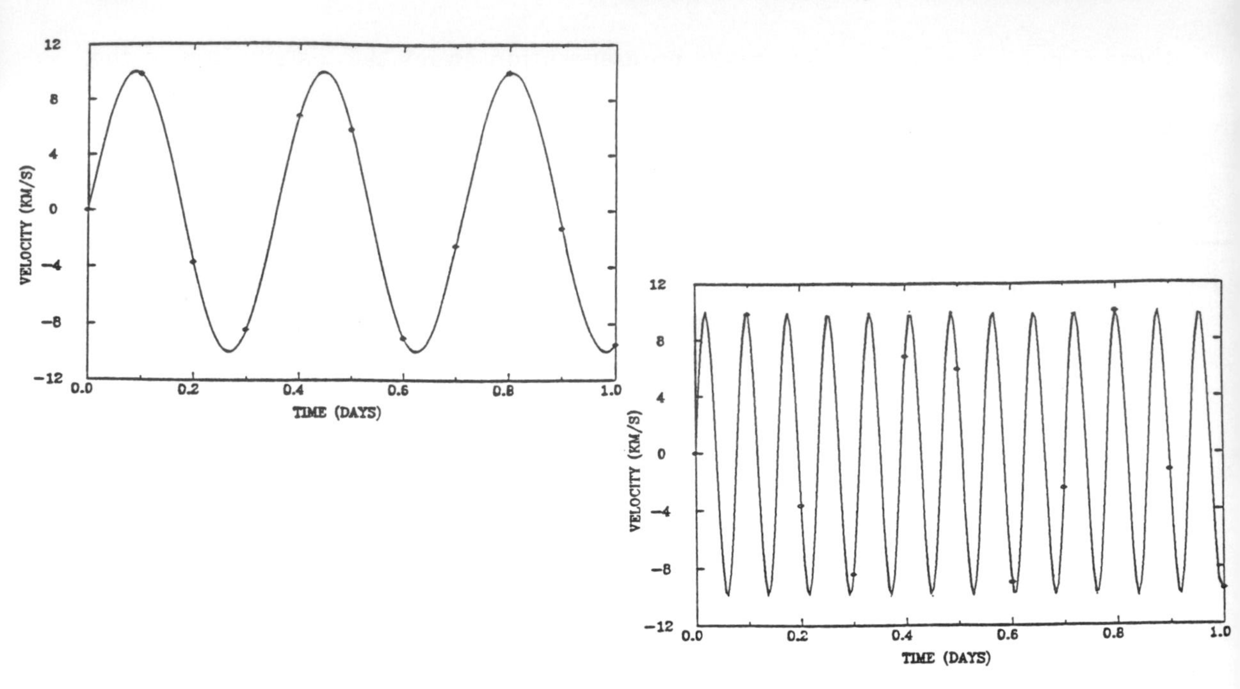

Figure 2. An illustration of aliasing. In this simulation, the data points (◇) were obtained at intervals separated by exactly 0.1 day, with "infinite" precision. (a) A sinusoid with amplitude 10 kms^{-1} and period 0.36 day (frequency 2.8 d^{-1}) fits the "observations" quite well. (b). However, the same data are fit equally well by a sinusoid with amplitude 10 kms^{-1} and period 0.08 day (frequency 12.8 d^{-1}). Thus the actual periodicity present in the data cannot be determined uniquely. The two possible periods are "aliases" of each other.

permit the calculation of a "window function" (*i.e.*, a periodogram of the independent variable). All information regarding the temporal sampling of a particular data set is contained in the window function (Deeming 1975; Scargle 1982). Comparison of the window function peaks with the pattern of pseudo-aliases in the periodogram of the data can aid identification of the true frequency, which is usually, though not always, the highest peak in the periodogram. Gray & Desikachary (1973) have formalized this procedure through their method of cross-correlating the window function with the periodogram. This approach utilizes all the information contained in the pattern of the pseudo-aliases, but it does not guarantee the successful identification of the genuine period. In most situations "prewhitening" can be used to determine which of a group of peaks corresponds to the physical period (see VI.1).

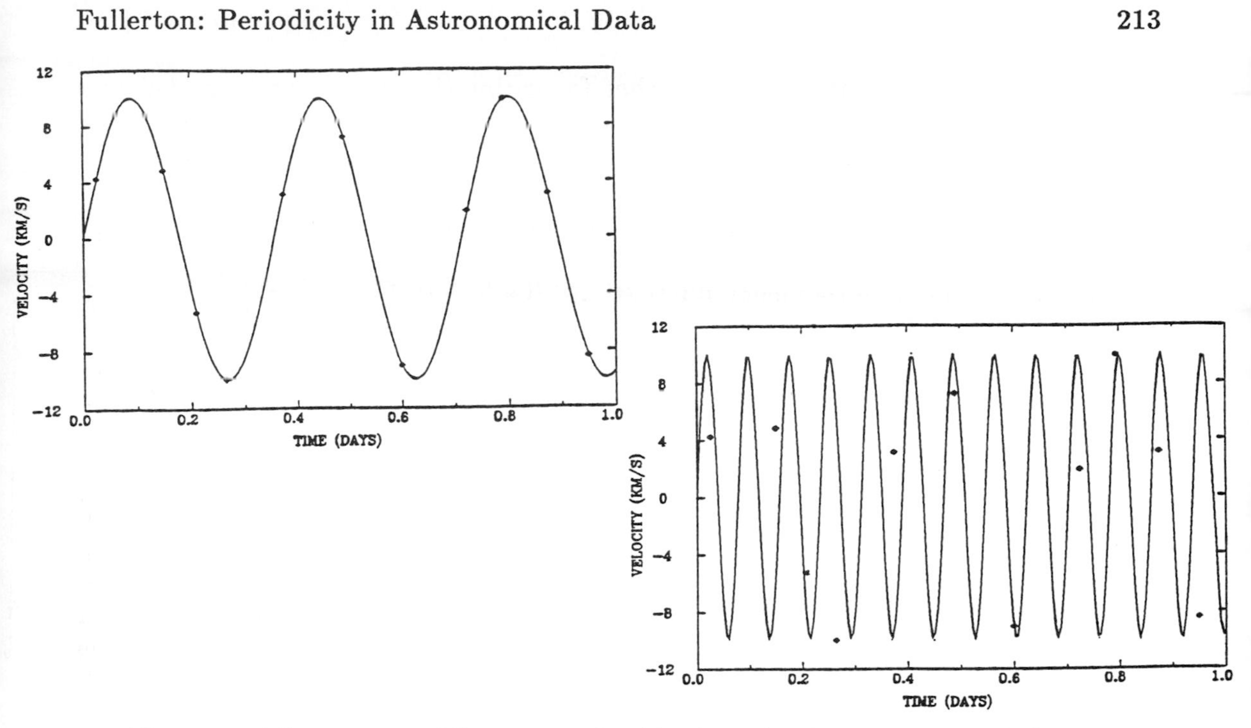

Figure 3. The same as Figure 2, only this time the data were sampled at unequally spaced times. The period of the variation is no longer ambiguous. Apparently the longer of the two periods illustrated in Figure 2 is the correct one.

V.2 Filling Gaps in the Data

Geophysicists frequently use the "maximum entropy method" (MEM) to fill temporal gaps in a data set, thereby preventing the occurrence of pseudo-aliasing. This technique uses the observations from one night as a template to link successive nights in a continuous and "maximally noncommittal way". Fahlman & Ulrych (1982) have demonstrated the power of this technique for astronomical applications. However, the usefulness of MEM suffers from its inability to utilize unequally spaced data as input. Nevertheless, MEM has very desirable frequency resolution characteristics, so even a single night's observations can yield good period estimates for stars which vary sufficiently rapidly (Percy 1977).

V.3 Noise Estimation

As discussed in section 4, the only protection against "detecting" spurious periods lies in rigorous statistical analysis. Accurate estimation of the noise level present in a data set is of considerable importance since the S/N ratio is the quantity by which the significance of a periodogram peak is assessed.

There are a variety of approaches for estimating the noise in a data set. The external precision of the observations (*e.g.*, the standard deviation of (Check-Comparison) observations in photometry or the mean error of a radial velocity measurement) provides a straightforward estimate. In principle the noise may be determined directly from a Fourier-type periodogram, under the assumption that the noise is locally white. This strong assumption is necessary because photometric observations suffer low frequency fluctuations due to hourly and nightly changes in atmospheric conditions. By averaging the periodogram over a region near the peak of interest (and preferably free of pseudo-aliases since these features are not random fluctuations) a good estimate of the mean square noise, σ_o^2, is usually obtained. Horne & Baliunas (1986) demonstrate that the total variance of the input data (in the time domain) is the *only* acceptable noise estimator for use with the Scargle periodogram if the valuable statistical properties of this algorithm are to be retained.

VI. SECOND ORDER COMPLICATIONS

In addition to the complications which plague attempts to recover single periodicities from a data set, further difficulties are associated with the analysis of multiperiodic or quasi-periodic phenomena.

VI.1 Prewhitening and the Detection of Multiple Periods

Prewhitening is a term used to indicate the removal of a signal at a particular frequency from a data set. Usually this is accomplished by fitting sinusoids of the appropriate frequency to the data by least squares. The residuals from the fit form the prewhitened data.

Prewhitening has two uses in time series analysis. The first, alluded to in section V.3), is to aid in the identification of a (single) "true" periodicity under conditions of unfavourable pseudo-aliasing. To decide which of the statistically significant peaks corresponds to the real frequency, the data are prewhitened for each of the aliased frequencies in turn and periodograms of the residuals are examined. When a pseudo-alias frequency has been removed the other peaks remain since a "physical" signal is still present. Often the periodogram of the residuals contains new peaks and the overall noise level rises. In contrast, prewhitening with the true signal frequency removes all the pseudo-alias peaks from the periodogram of the residuals; the overall noise level usually decreases. Thus the true frequency may be distinguished from the pseudo-aliases in many cases.

More frequently prewhitening is used to search a data set for more than one periodicity. This procedure involves the detection of a period through periodogram analysis, its subsequent removal by prewhitening, and reanalysis of the residual data for more periods. In principle this step-by-step approach is valid for analyzing a linear combination of sinusoidal, multi-periodic signals. In practice the detection of more than one period is very difficult, primarily because the prewhitening procedure destroys the independence of the input data. This is particularly true if the prewhitening fits possess large uncertainties. As a result, the statistical significances of the peaks in the periodogram of the prewhitened residuals are impossible to evaluate analytically.

This bleak outlook is corroborated by a series of tests on synthetic data performed by Walker, Pike, & Hartley (1984). Their conclusions are worth repeating:

1) "It is well known that any noisy data set can be adequately (*i.e.*, down to the noise level) modelled by a power spectrum of sufficient complexity. Typical large data sets on a variety of stars often require 10 frequencies.

2) The power spectrum for typical data sets is not unique and if the wrong frequency is used at any stage in the complex fitting and prewhitening procedure then a different frequency set will be found which will fit the data equally well.

3) A typical data set is so poorly constrained that the total number of frequency sets which will adequately represent the data is large and may be infinite. Alternately we might summarize by saying that generally one needs a lot more data than one originally thought...."

Often, no more than one periodicity can be reliably extracted from a data set.

VI.2 Detection of quasi-periods (timescales)

The period search techniques discussed in this paper are valid for "strictly" periodic signals only (see section II.3). However, there are numerous phenomena in astronomy which are not periodic but which do change over characteristic time intervals (*e.g.*, photometric variations of supergiants, Be stars, spotted rotating stars, quasars). These timescales convey useful physical information about the origins of the variability, and are therefore worthy of detailed study.

Autocorrelation methods have been used successfully to estimate timescales (Burki, Maeder, & Rufener 1978; Percy, Jakate, & Matthews 1981; Baliunas *et al.* 1983). These techniques differ from those discussed in section III in that the analysis occurs in the time domain. All autocorrelation methods seek to determine the time lag required for the observations to recognize or report themselves. This is accomplished by introducing a series of discrete time shifts to the data. At each step in the procedure the shifted data are compared with the unshifted data by means of a "correlation function". A simple example of such a function is just the sum of the squared differences between the parts of the shifted and unshifted data which overlap. A plot of the correlation function *vs.* time delay has a large dip when the data sets are in approximate registration, *i.e.*, when the shifted data "recognizes" itself in the unshifted data. This value of the time delay corresponds to the timescale, or at least to a harmonic or multiple of the actual timescale of the variation.

As with the frequency domain analysis of strictly periodic phenomena, statistical significances must be calculated for the dips observed in autocorrelation diagrams. Baliunas *et al.* (1983) discuss a method of calculating statistical significances for their autocorrelation technique. Heck, Manfroid, & Mersch (1985) examined the performance of an autocorrelation algorithm in their extensive series of tests of period search methods. As expected, they found that it was not as reliable as the other methods for estimating strictly periodic signals. However, they stress that the real utility of the autocorrelation method lies with the problem of determining quasi-periods.

VII. Concluding Remarks

In their famous treatise on mathematical methods, Jeffreys & Jeffreys (1953) make the following remarks on period searching: "Without some such precaution periodicities found by harmonic analysis and not predicted by previous theoretical considerations should be mistrusted, as many complications are capable of giving rise to spurious periods; not more than a tenth of those that have been asserted will bear a proper statistical examination."

The theme of this paper has been to echo these sentiments. In particular, the distinction between detection ("periodicities found by harmonic analysis") and estimation ("predicted by theoretical considerations", or by independent observations) has been emphasized. The importance of rigorous statistics, particularly for the detection of suspected signals, has also been stressed, primarily to improve upon the ten per cent success rate suggested by Jeffreys & Jeffreys!

Even with rigorous statistical analysis, the complexities associated with period detection and estimation can sometimes confound a researcher. In such cases only a thorough knowledge of the behaviour of the period search method used will prevent the proliferation of spurious periods in the literature. Thus, every method should be tested comprehensively using both simulated and real data. Careful, conservative analysis of "detection" periodograms will eliminate a great many spurious periods, but, in the final analysis, further observations will always tell the tale!

Acknowledgements

It is with great pleasure that I record my gratitude to Drs. C. T. Bolton and J. R. Percy for their encouragement and instruction in learning the black art of period searching. Many "periodic" discussions with Dr. D. R. Gies, C. R. Stagg, and M. F. Bietenholz have been extremely valuable to the formulation of my thoughts, as have comments from D. L. Welch and L. M. Oattes. Finally, financial support from the Natural Sciences and Engineering Research Council of Canada is gratefully acknowledged.

REFERENCES

Baliunas, S. L. *et al.* . (1983) *Ap. J.*, **275**, 752.
Black, D. C. & Scargle, J. D. (1982). *Ap. J.*, **263**, 854.
Burki, G., Maeder, A., & Rufener, F. (1978). *Astr. Ap.*, **65**, 363.
Deeming, T. J. (1975). *Ap. Space Sci.*, **36**, 137.
Dworetsky, M. M. (1983). *M. N. R. A. S.*, **203**, 917.
Fahlman, G. G. & Ulrych, T. J. (1982). *M. N. R. A. S.*, **199**, 53.
Gray, D. F. & Desikachary, K. (1973). *Ap. J.*, **181**, 523.
Heck, A., Manfroid, J., & Mersch, G. (1985). *Astr. Ap. Suppl.*, **59**, 63.
Horne, J. H. & Baliunas, S. L. (1986). *Ap. J.*, **302**, in press.
Jeffreys, H. & Jeffreys, B. S. (1953). *Methods of Mathematical Physics*, 3rd edition. p. 452. London: Cambridge University.
Jensen, O. G. & Ulrych, T. J. (1973). *A. J.*, **78**, 1104.
Jurkevich, I. (1971). *Ap. Space Sci.*, **13**, 154.
Kurtz, D. W. (1983). *Inf. Bull. Var. Stars*, # 2285.
Lafler, J. & Kinman, T. D. (1964). *Ap. J. Suppl.*, **11**, 216.
Lomb, N. R. (1976). *Ap. Space Sci.*, **39**, 447.
Middleditch, J. & Nelson, J. (1973). *Astrophys. Lett.*, **14**, 124.
Moffett, T. J. & Barnes, T. G. (1985). *Ap. J. Suppl.*, **58**, 843.
Morbey, C. L. (1973). *Pub. D. A. O.*, **14**, 185.
Morbey, C. L. (1978). *Pub. D. A. O.*, **15**, 105.
Nemec, A. F. L. & Nemec, J. M. (1985). *A. J.*, **90**, 2317.
Percy, J. R. (1977). *M. N. R. A. S.*, **181**, 647.
Percy, J. R., Jakate, S. M., & Matthews, J. M. (1981). *A. J.*, **86**, 53.
Perez de la Blanca, N. & Garrido, R. (1981). In *Workshop on Pulsating B Stars*, eds. G.E.V.O.N. & C. Sterken, p. 285. Nice: Nice Observatory.
Ponman, T. (1981). *M. N. R. A. S.*, **196**, 583.
Priestly, M. B. (1981). *Spectral Analysis and Time Series*, Chapter 6, London: Academic.
Renson, P. (1978). *Astr. Ap.*, **63**, 125.
Scargle, J. D. (1982). *Ap. J.*, **263**, 835.

Stellingwerf, R. F. (1978). *Ap. J.*, **224**, 953.
Vanicek, P. (1969). *Ap. Space Sci.*, **4**, 387.
Vanicek, P. (1971). *Ap. Space Sci.*, **12**, 10.
Walker, E. N., Pike, C. D., & Hartley, K. F. (1984). In *Space Research Prospects in Stellar Activity and Variability*, eds. A. Mangeney and F. Praderie, p. 151. Paris: Observatoire de Paris-Meudon.
Wizinowich, P. & Percy, J. R. (1979). *Pub. A. S. P.*, **91**, 53.

THE O–C DIAGRAM: A USEFUL TOOL

L. A. Willson
Astronomy Program
Physics Department
Iowa State University
Ames, IA 50011
USA

Observations of stars are made for two reasons: to check whether our ideas about a star or star system are correct, and to look for "surprises". The $O - C$ diagram is a tool which can be used both to check whether the light curve of a star is behaving as we expect, and to look for more subtle variations than we can find by simply inspecting the light curve.

By the "$O - C$ for maximum light", for instance, we mean simply the **O**bserved time of maximum minus the **C**alculated time of maximum based on a known or estimated period (P) and epoch (E):

$$C = E + nP \tag{1}$$

The "$O - C$ diagram" is a plot of $O - C$ *vs.* time or, more commonly, *vs.* cycle number, n. Thus if we have observed maxima at times $(t_o, t_1, t_2, \ldots, t_N)$ and know the period P we can find

$$(O - C)_n = t_n - (t_o + nP). \tag{2}$$

Here, n should always be chosen such that $(O - C)_n < P$ — that is, we need to use the observed and calculated times of maximum for the same cycle.

To form the $O - C$ diagram we must have an estimate for the period. There are many techniques for determining the period(s) of a variable star: Fourier techniques, Maximum Entropy methods, "stacking" techniques such as Phase Dispersion Minimization, or just plain trial-and-error. These are described in the paper by Alex Fullerton in this volume, and will not be discussed here.

The $O - C$ diagram is a very useful technique for checking (a) the accuracy of the period derived by other techniques and (b) whether the star has a single, constant period. Most other techniques don't tell us whether the period is constant, for the simple reason that they generally start by assuming that it is.

If the estimated period is in fact the correct period, then the $O - C$ diagram will consist of a scattering of points around a horizontal line. If the epoch is in error by some amount δE, the average value $\langle O-C \rangle = \delta E$ provides a ready correction. If the initial period P_{est} is wrong, but P is constant, the plot will be a straight line: $O-C = n(P - P_{est})$. In the $O-C$ diagram, this means that if P_{est} is too long, the $O-C$ line will slope downwards. The accuracy to which the period can be determined in this way depends on the scatter (σ) in the points and on the number of cycles observed, N; the uncertainty in P is $\pm(\sigma/N)$.

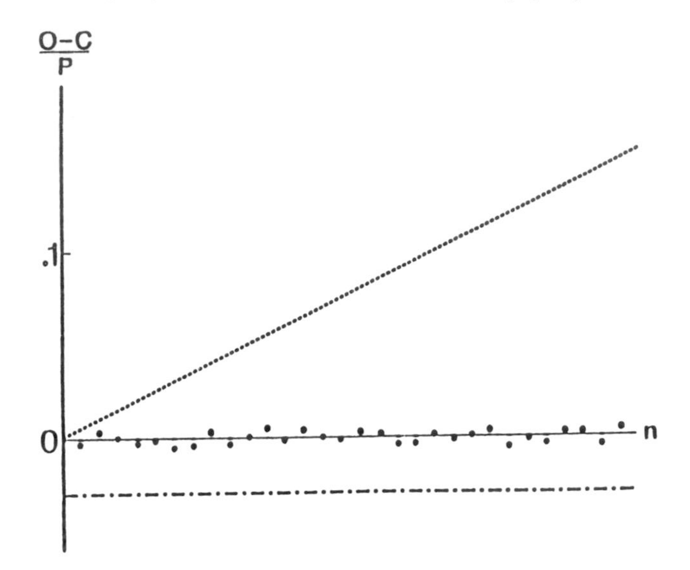

Figure 1. Characteristic $O-C$ diagram for the case of a constant period. The dots along the horizontal axis show the typical behaviour of the $O-C$ for a constant, correct period. The dashed line shows the $O-C$ when the assumed period, P_{est}, is too short; $O-C = n(P-P_{est}) = n(.005P) = 0.1$ when $n = 20$. The dot-dash line indicates $O-C$ for an incorrect epoch: the true $E = E_{est} + 0.03P$.

While the $O-C$ diagram provides a ready means of improving the accuracy of the period and epoch for a variable star, its most important function is as a detector of period and phase changes. All common period search techniques assume that the period(s) are constant over the interval of the observations, and that the phases are stable. These conditions are not always met by variable stars. It is important to use an $O-C$ diagram to check whether the conditions of period and phase stability are met, as part of the verification of the results of any period search technique, whenever the light curve is sufficiently well observed to make the construction of an $O-C$ diagram possible.

A change in the phase of variation is essentially a change in the epoch, E; it therefore shows up as a vertical shift of the $O-C$ line (Figure 2a). More common than phase changes are period changes; these give changes in the slope of the $O-C$ line. If the period increases abruptly, the old period will be too short, and the $O - C$ line will bend upwards; if the period decreases abruptly, the line will bend downwards. Abrupt period changes are often observed in RR Lyrae stars (the reason is not yet clear) and in binary star systems when mass is abruptly lost by one of the component stars.

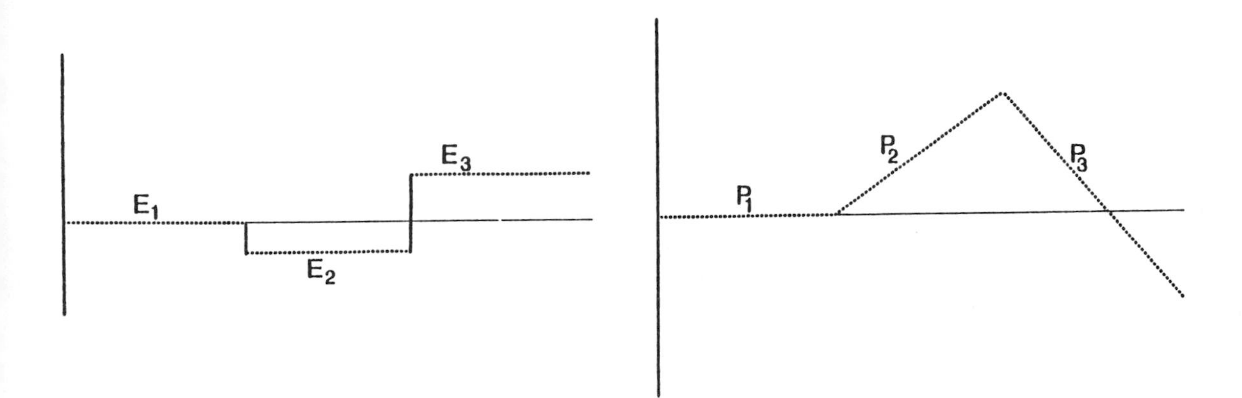

Figure 2. The effects on the $O - C$ diagram of abrupt changes in epoch and period. For the example shown, $E_2 > E_1 > E_3$ and $P_2 > P_1 > P_3$.

If the period is slowly changing — for example, as the stellar radius or mass changes due to evolutionary effects or mass loss — then the $O - C$ diagram will be curved:

$$O - C = n(P(t) - P_o). \tag{3}$$

In particular, the $O - C$ curve will be parabolic if the rate of period change is constant, so that the change in P per cycle, $\Delta P/P$, is constant:

$$(O - C)_n = (\Delta P/P)n^2 \tag{4}$$

If we take P_o and E to be the correct period and epoch at the time of the first observed maximum, then:

$$(O - C)_n = (\Delta P/P)(n - n_o)^2 + (E(n_o) - E) \tag{5}$$

if P_o is correct at cycle n_o and E is the epoch used in C.

To fit a parabolic $O-C$ and thus to derive $(\Delta P/P)$ one can use standard least squares techniques — which essentially minimize the scatter about the parabola — or one can proceed graphically. In the second case it is probably easiest to use the average period over the interval of the observations for P_o; then the $O-C$ parabola will have its maximum or minimum value at the centre of the observing interval (Figure 3), and we can take advantage of the symmetry to check that the parabola is a good fit.

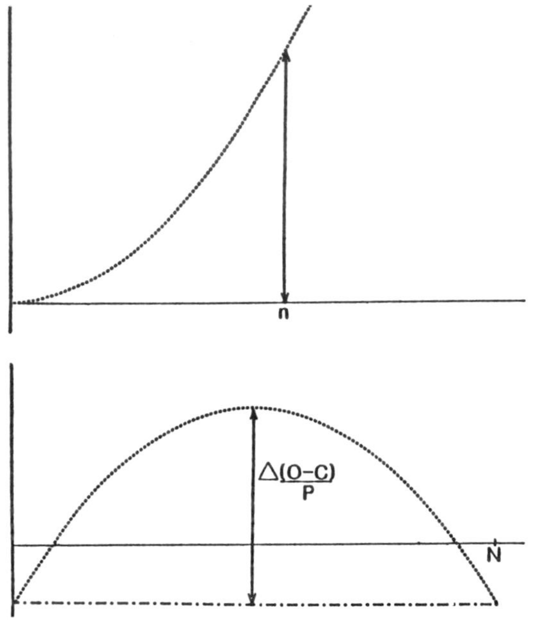

Figure 3. Parabolic $O-C$ diagrams for constant $(\Delta P/P)$. Top: plotted with $P_{est} = P(n=0)$; bottom: plotted with C calculated for the average period during the interval $n = 0$ to N. In the upper plot, the deviation of $O-C$ from zero at cycle n is $(\Delta P/P)n^2$. In the lower plot, $\Delta(O-C)/P = (\Delta P/P)N^2/4$.

The accuracy with which period changes can be measured for a given star increases as n^2; thus if the star is observed for twice as long, period changes 1/4 as large can be determined. It can therefore be more important to extend the observations for a well-studied system than to observe a new one!

Analyses of average $\langle \Delta P/P \rangle$'s by this technique have detected evolutionary changes for Cepheids, RR Lyrae stars, and selected Miras; such observations provide constraints on stellar evolution theory unmatched by any other techniques. Also, for binary systems, $O-C$ analyses of eclipses can reveal very gradual changes due to tidal interactions or even, for close binaries of degenerate stars, due to gravitational radiation.

One cause of *apparent* period changes can be the presence of a secondary period in the data. Here, again, the $O - C$ diagram can be a useful part of the analysis. When a secondary period is present, the result will be a periodic variation in $O - C$. The precise form of the $O - C$, and the relation between the $O - C$ period and the secondary period in the light curve, depend on the kind of secondary period which is present.

Perhaps the simplest case of a secondary period is the case of amplitude modulation. Amplitude modulations (as in the AM radio signal) give light curves of the form

$$L(t) = L_o + A(t)\sin(2\pi t/P) \tag{6}$$

where, for example, $A(t) = \sin(2\pi t/P_A)$ for a sinusiodal modulation. Amplitude modulation generally gives very small or undetectable variations in the $O - C$ diagram: the observed time of maximum is only very slightly altered by the small change in $A(t)$ near the time of maximum light of each cycle if the modulation period (P_A) is much longer than the cycle length. Amplitude modulation can occur as the result of internal processes in a pulsating star, and is observed for some objects.

In contrast, frequency modulation (as for FM radio) — which is equivalent to period modulation — gives large variations in $O - C$. For frequency modulation,

$$L(t) = L_o + A\sin(2\pi t/P(t)) \tag{7}$$

and times of maximum occur when $(2\pi t/P(t)) = (2n + 1)\pi/2$. Frequency modulation may occur, for example, when a pulsating star is in a wide binary system: the extra time which light takes to reach us from the far side of the star's orbit compared to the time when the star is closer can be a significant fraction of the period of the star. Equivalently, the period can be thought of as changing because of the Doppler effect as the star moves towards or away from us.

Perhaps the most common, and the messiest, case of multiple periods for variable stars occurs when there are two or more pulsation periods excited simultaneously. For two simultaneous small amplitude modes in a pulsating star the brightness variation will typically be of the form

$$L(t) = L_o + A\sin(2\pi t/P_1) + B\sin(2\pi t/P_2 + \phi) \tag{8}$$

where ϕ is the difference in phase between the two modes at time $t = 0$.

Maxima (or minima) in the light curve will occur when the slope of L, dL/dt, is zero; application of standard calculus to (8) and assuming that $(O - C) \ll P_1$, for C calculated for P_1, gives (Peterson 1980)

$$O - C = \frac{-P_2}{2\pi} \frac{\sin(\frac{2\pi C}{P_2} + \phi)}{(\frac{P_2}{P_1})^2 \frac{A_1}{A_2} + \cos(\frac{2\pi C}{P_2} + \phi)} \tag{9}$$

which has a period of P_2 in time units or P_2/P_1 in units of n, the cycle number, if $P_2 \gg P_1$. More typically, however, P_2 and P_1 are close — within a factor of 2 or 3; in that case, the $O-C$ will vary with the beat period $P_b = \pm(1/P_2 - 1/P_1)^{-1}$. In the particular case where the two amplitudes are about the same and the periods are close, the $O-C$ diagram will have a saw-toothed appearance as the $O-C$ will change abruptly by $0.5P_1$ at the minimum of the modulation cycle.

From equation (9) we can find the amplitude of the $O-C$ curve: $(O-C)/P_1 = P_1A_2/2\pi P_2A_1$ if $(P_2/P_1)^2(A_1/A_2) \gg 1$ (as is frequently the case). This implies that for a given A_1 and A_2, longer P_2 will give a smaller $O-C$ amplitude. A_1 and A_2 are readily deduced from the light curve; thus P_2 can be unambiguously determined. In this simple case, P_2 could be equally unambiguously determined from an inspection of the light curve, which appears quite different in the two cases; in more complicated situations this is not always the case, and the $O-C$ can be used to resolve the ambiguity (Peterson 1980).

Other kinds of sinusoidal modulations are also possible, and will show up in the $O-C$ diagram. In Figure 4 (reproduced from Peterson 1980) two apparently similar light curves have been computed from very different assumptions; however, the $O-C$ amplitudes are very different, and allow the two cases to be readily distinguished. As a general rule of thumb, amplitude modulation (Eq. 6) or the interaction of two very dissimilar periods ($P_1 \gg P_2$ in Eq. 8) lead to smaller $O-C$ amplitudes than do frequency modulation (Eq. 7) or the beating of two close periods ($P_1 \sim P_2$ in Eq. 8). In analyzing the variations of a star with a periodic $O-C$, it is important to consider several hypotheses about the nature of the variation, and to construct both synthetic $O-C$ diagrams and synthetic light curves for each possibility.

Some pitfalls in the construction and interpretation of $O-C$ diagrams need to be mentioned; as for any data analysis method, the care with which it is applied is an important factor in determining the validity of the conclusions which are drawn.

First, what can be learned from an $O-C$ diagram depends to some degree on the choice of the phase to be plotted. Maximum light is often used for pulsating stars: large amplitude stars are more readily and accurately observed when they are brightest. The effects of the beating of two close frequencies will be particularly obvious in the timing of maximum light, giving a large (and hence relatively accurately determined) $O-C$ amplitude. However, maximum is also the time when the rate of change of brightness of the star is minimum, and this can lead to large uncertainty in the observed time of maximum. For this reason, some astronomers prefer $O-C$ diagrams for rising light, where the time at which the light curve crosses some reference brightness can be very accurately determined. For eclipsing

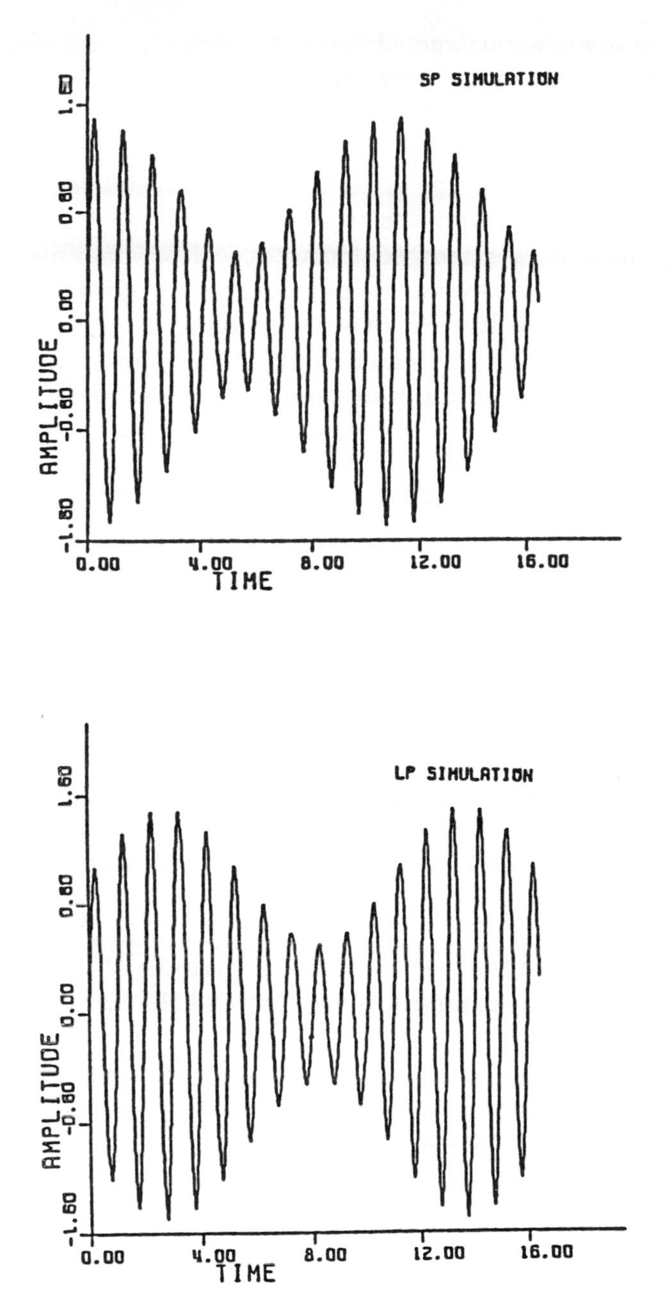

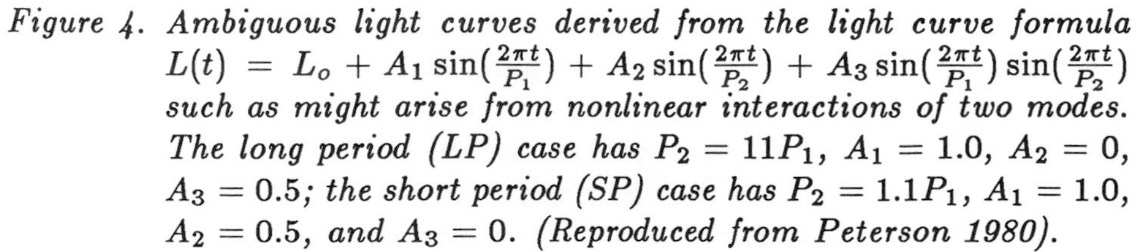
Figure 4. Ambiguous light curves derived from the light curve formula $L(t) = L_o + A_1 \sin(\frac{2\pi t}{P_1}) + A_2 \sin(\frac{2\pi t}{P_2}) + A_3 \sin(\frac{2\pi t}{P_1}) \sin(\frac{2\pi t}{P_2})$ *such as might arise from nonlinear interactions of two modes. The long period (LP) case has* $P_2 = 11P_1$, $A_1 = 1.0$, $A_2 = 0$, $A_3 = 0.5$; *the short period (SP) case has* $P_2 = 1.1P_1$, $A_1 = 1.0$, $A_2 = 0.5$, *and* $A_3 = 0$. *(Reproduced from Peterson 1980).*

binaries, the $O-C$ is always constructed from minimum light, because that is the time which it is easiest to determine very accurately.

The information revealed by the $O-C$ depends also on the method used for deriving O from the data. For noisy light curves, several techniques are used to determine the time of maximum: the time of the brightest single observation; the maximum brightness in a smoothed light curve (for example, a light curve which includes the averages of several observations); the phase of maximum of a representative mean light curve fitted to all or a portion of the observations over that cycle; the maximum derived from an exact solution for a Fourier series which has been fitted to the data; or Hertzsprung's method of extrapolated bisectors. Several of these are illustrated in Figure 5 for a representative noisy and asymmetric light curve such as might be obtained for a faint long period variable. These methods all give slightly different results, as can be seen in this example. The best method to use depends on the characteristics of the observed light curve and on what information is being sought from the analysis. For example, the fitting of a representative average light curve to the observations has the advantages that it (1) makes use of all the observations, not just the one or several brightest ones, and (2) it suppresses "noise" due to cycle-to-cycle variations, hence gives the best $O-C$'s for searching for long term changes in P or E. (This is the method used by the AAVSO in determining the maxima of long period variables.) However, this method also decreases the $O-C$ amplitude from the beating of two close periods, so is a less appropriate method to use when seeking closely spaced multiple periods. It is important, of course, to construct each $O-C$ diagram with observed maxima derived by a consistent method.

Another caution, particularly for the case of periodic $(O-C)$'s: the relation between the type of variation (beating, AM, or FM) and the amplitude of the $O-C$ is very sensitive to the shape of the underlying variation. A saw-tooth light curve with an amplitude modulation will show an absolutely flat $O-C$, while a sinusoidal light curve with the same modulation may have a significant variation in $O-C$, depending on the method used for determining O. Before making use of the $O-C$ amplitude information it is important to find an analytic fit (*e.g.*, a Fourier series) to the mean light curve, and to use that fit to predict the $O-C$ amplitudes for the cases to be tested.

Finally, in any analysis of light curves it is important to consider the distinction between the magnitude curve and the luminosity or brightness curve. Since magnitudes are logarithmic quantities, they compress big numbers and expand small ones. The determination of the mean period will not generally depend on the choice of magnitude *vs.* luminosity, but the other information to be derived from the $O-C$ diagram may. If the amplitude of variation is sufficiently small (much less than 1

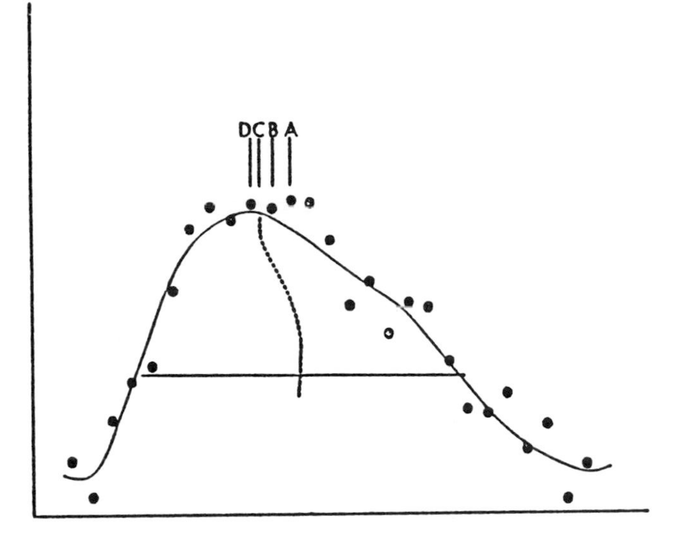

Figure 5. An asymmetric and noisy light curve, and the "observed maxima" derived by (A) locating the brightest single point, (B) locating the peak in a smoothed version obtained by averaging sets of 3 neighbouring observations, (C) extrapolating the bisectors of the light curve (indicated by the dashed line — Hertzsprung's method), and (D) fitting a mean light curve (solid line) and using the maximum of the mean light curve.

magnitude) then the magnitude curve and the brightness curve will be quite similar. However, for amplitudes of about a magnitude or more, the magnitude and the brightness curves may give very different results!

The $O - C$ diagram is one example of a technique which is easy to use, and which converts data into a form which is well suited to analysis by humans (whose ability to recognize patterns and anomalies vastly exceeds that of electronic computers). Other techniques in the same class include the use of histograms, plots of data in different formats, and various smoothing techniques. For the serious data analyst, the book *Exploratory Data Analysis* by J. W. Tukey (originator of the widely used "Fast Fourier Transform") contains a wealth of suggestions.

An acknowledgement is due to Dan Peterson, whose thesis work on the $O-C$ amplitude test brought some of the subtler aspects of $O-C$ diagrams to the author's attention. J. R. Percy's encouragement is also appreciated, as well as the hospitality of the University of Toronto Astronomy Department and the Canadian Institute for Theoretical Astrophysics during the author's 1985-86 Faculty Improvement Leave from Iowa State University.

REFERENCES

Peterson, D. (1980). *Ph.D. Thesis*, Iowa State University.

Tukey, J.W. (1977). *Exploratory Data Analysis.* Cambridge, Mass: Addison-Wesley.

FOURIER TECHNIQUES FOR CLOUDY NIGHTS

Emilia Pisani Belserene
Maria Mitchell Observatory
3 Vestal Street
Nantucket, MA 02554
U.S.A.

What can a variable star observer do *on* cloudy nights? What can the variable star researcher do *about* cloudy nights? Fourier techniques are an ideal solution to both problems, if the small telescope is supported by a small computer.

A complete Fourier Transform cannot be done. It would require continuous and infinite data. The famous Fast Fourier Transform is also ruled out. It requires evenly spaced data, with no interruptions by clouds. There is a Discrete Fourier Transform by Deeming (1975), which allows the data to be neither continuous nor evenly spaced. Even better is the Data-Compensated Discrete Fourier Transform (DCDFT) due to Ferraz-Mello (1981). It fits a four-parameter sinusoid to irregularly spaced data, and allows for the possibility that the mean of the data does not necessarily coincide with the zero level of the sinusoid. The four parameters are: in magnitude: zero-point and amplitude; in the time coordinate: period (or frequency) and epoch ("phase" in Fourier jargon). In other words, one gets the equation of a sinusoidal approximation to the light curve, and the linear elements (epoch of maximum and period) which correspond to this approximation. The sinusoid can be filtered out, and the residuals searched for other periods that may be in the data. The filtering step also removes aliases of the main period that may be in a Fourier transform due to the distribution of the dates. Secondary periods in the data may be masked by these aliases so that they would not show up in a transform of the original data; they may show up strongly in the residuals.

What about the fact that a light curve is not necessarily sinusoidal? Another Fourier technique can handle that problem. Any periodic function can be approximated by the sum of several terms of a Fourier series, involving submultiples of the basic period. It is easy to forget that these submultiples are not additional periods allowed by the physics. They simply describe the deviation from a simple sinusoid. If they have significant amplitude they must be removed, or else they and their aliases may look like additional modes of pulsation of the star.

At the Maria Mitchell Observatory we have written this procedure in FORTRAN for use on an IBM PC or remotely on a VAX at the Harvard-Smithsonian Center for Astrophysics. We also have a less-precise version that fits into a little TRS-80 Model I.

The strategy is to use the DCDFT for period search, using a coarse grid of trial periods. Then the vicinity of the most promising period or periods can be scrutinized with a finer grid. The next step is to use Ferraz-Mello's harmonic filter to remove this best sinusoid, namely, the first term of a Fourier series with this period. Its aliases are removed at the same time, *i.e.*, the data set is "whitened" and the residuals will have a better chance to show any additional real periodicities.

The next piece of strategy is to look at the 2nd and higher terms of the Fourier series, in case the light curve is not well approximated by a sinusoid. Probably a Fourier series program (*e.g.*, Gautschy 1982) would be a better way to do this, but additional runs through the DCDFT serve the purpose pretty well. It is informative to watch the sum of the squares of the residuals go down as any significant periods are filtered out.

The final bit of strategy is to search the residuals for any additional periods in the data. We have had success with this program on DF Cyg, an RV Tauri star whose 50-day cycle is superimposed on a 770-day variation in mean magnitude. We hope to use it on RR Lyrae stars with the Blazhko effect. At the time of this writing, the only second period it has picked up turned out to be an alias of a submultiple of the fundamental period. That this apparent period disappeared when the corresponding submultiple was filtered out gives us confidence that the procedure works as intended.

The question always arises whether a period seen in data is likely to be real or simply an artifact of noise in the data. Our program includes an analysis of the statistical significance of the best period, following Scargle (1982). His method is a way of finding the probability that the search would have turned up as nice a result even if the data contained only noise. A problem is that the method does not apply rigorously to the DCDFT. It seems to me, however, to give a useful estimate. A second problem is that it requires knowing how many of the periods searched are statistically independent. The Maria Mitchell version includes several hypotheses as to the number of independent periods. For example, if the observations are evenly spaced in time the number of independent trials is half the number of observations. This is one of the possible hypotheses. Preliminary trials with constructed data, however, suggest that this value is too low when the time intervals are irregular, so hypotheses based on the typical time intervals between the observations (as if there were no gaps due to clouds) are included. An interesting special case occurs when there is only a single trial period chosen *a priori*, perhaps because it was suggested by theory. Then the number of independent periods is just one. Scargle emphasizes that this is not the case when the period has been found from the analysis. There is a statistical penalty to be paid when choosing the best of a number of trials. The statistical section of our DCDFT leaves to the user to decide how serious this penalty is.

I am grateful to the National Science Foundation for a grant, AST 83-20491, which has materially improved our computing capability. We are willing to share the code with others who would like to try the process.

REFERENCES

Deeming, T.J. (1975). *Ap. Space Sci.*, **36**, 137.
Ferraz-Mello, S. (1981). *A. J.*, **86**, 619.
Gautschy, A. (1982). *IAPPP Comm.*, **10**, 11.
Scargle, J.D. (1982). *Ap. J.*, **263**, 835.

VII. NEW DIRECTIONS

In what directions will the study of variable stars using small telescopes go in the future? To answer such a question, it is often helpful to look to the past. This was a reason for asking Russell M. Genet to speak in a historical rather than technical vein, and for including his paper in this section of these proceedings. He is a "textbook example" of an amateur astronomer. With professional competence, he has applied his particular skills as a human factors engineer to the task of doing astronomical photometry in an automated way. His review certainly illustrates how both human and technological factors have influenced the development of this field.

Photoelectric photometry with small telescopes became simple and inexpensive in the 1970's. Automated photometry became a reality a few years later. Future developments will be based on the available technology and on the premise that astronomers, whether amateur or professional, will do what is interesting and worthwhile. For instance, when photoelectric photometry is done routinely by automated telescopes, "live" astronomers may concentrate on adapting and using new technology (such as CCD's) in small-telescope astronomy, as they have done in the recent past. They may become more involved in infrared photometry or in spectroscopy, or in data analysis and interpretation. If they choose to continue to do broadband photometry, they may concentrate on observing stars which are too faint, crowded or otherwise unsuited for automated photometry. Or they may simply take advantage of "Argelander's Law": there will always be more variable stars than observers.

These proceedings end with one astronomer's view of what the future of small-telescope astronomy may hold. Slavek M. Rucinski was asked to speak on this topic because he has an imaginative mind, an appreciation of modern technology, and years of experience in studying binary and variable stars with a variety of ground-based and space-based instruments. The success of his presentation is illustrated by the fact that, after it, he and Russ Genet were hard at work designing (at least in the mind) an automated spectroscopic telescope!

SMALL AUTOMATIC PHOTOELECTRIC TELESCOPES: A HISTORICAL REVIEW

Russell M. Genet
Fairborn Observatory
Automatic Photoelectric Telescope Service
1357 N. 91st Place
Mesa, Arizona 85207
USA

I. INTRODUCTION

The assigned objective of this chapter was the review of automatic photoelectric telescopes. Webster's dictionary disclosed that a "review" was a "military parade" — obviously not appropriate — or a "historical summary" — certainly more so! The dictionary suggested that "automatic" (machines) were devices that ran themselves, replacing human senses and motions with electronic and mechanical devices. "Photoelectric" (or more properly "photometric") suggests the dedicated measurement of light intensity or colour (of variable stars) by electrical means. "Photoelectric" in this case is also meant to suggest a permanently mounted photometer that is also used to find and centre the stars to be measured. "Small" telescopes will be arbitrarily defined as those ground-based telescopes that are roughly 12 inches (30 cm) or less in diameter, *i.e.*, the size typically found at small college or backyard observatories.

Thus this paper will provide a historical review of small ground-based telescopes operating by mid 1985 with permanently mounted photometers that have been used to find, centre, and measure variable stars without human help. With these restrictions, it is only necessary to discuss four automatic photoelectric telescopes (APT's) — which makes the task relatively straightforward and the result (hopefully) within the allotted space. I apologize for not including APT's larger than the arbitrary limit, space-based APT's, APT's that look at Polaris and do not require photometric search and centering, APT's at the south pole that observe at a constant zenith angle and also do not require search and centering, and APT's under development that have not yet observed variable stars.

A special note of thanks is due to John Percy who invited and encouraged this chapter; to Arthur Code who supplied most of the material presented here on the Wisconsin APT; to David Skillman who provided a telephone interview about his APT; to Louis Boyd — my close associate for five years — for providing the early history of his system; and to Louis Boyd, Frank Melsheimer, John Diebold, Gerry Persha, and Lloyd Slonaker for their help in developing the second generation Fairborn APT.

II. THE WISCONSIN APT

It is certainly fitting that the first APT was developed at the University of Wisconsin, for it was here that Joel Stebbins, the father of photoelectric photometry, did most of his work. It was also here that Gerald Kron received his training. Kron later (at Lick Observatory) introduced the classic photomultiplier/DC amplifier/stripchart recorder combination. The authorative paper on the Wisconsin APT project is that of McNall, Miedaner, & Code (1968), and remains well worth reading by all those interested in APT's. In this chapter we will consider the more informal, historical aspects of the Wisconsin APT project. This is based on notes kindly supplied by Arthur Code.

The development of the Wisconsin APT can not be clearly separated from the activities centered around the construction of the Wisconsin Experimental Package for the Orbiting Astronomical Observatory (OAO), which started in the spring of 1958. Arthur Code responded to a request by the National Academy of Sciences, asking what could be done with a 100 pound satellite. Later that year NASA was founded, and the University of Wisconsin received funds for a study. The Space Astronomy Laboratory at the University of Wisconsin came into existence formally on June 16, 1959. Besides Code, the Space Astronomy Laboratory consisted of five graduate students working as project assistants, a draftsman and secretary, as well as Don Osterbrock and Ted Houck as consultants. Two of those graduate students are now intimately involved in Space Telescope (Bob O'Dell and Dan Schroeder). By July of 1959, Code and Houck had designed a 100 pound satellite which used 8 inch telescopes with off-axis photometers. It was this early design that eventually became four of the seven instruments in OAO and also the photometer/telescope mounted on the Wisconsin ground-based APT. The basic idea of an analogue and digital amplifier for this instrument was developed by Don Taylor, a graduate student at that time.

An AC Spark Plug Titan Missile Guidance System was picked up at a local surplus outlet between 1960 and 1962. A visiting astronomer, Dale Vrabec, was amazed that Wisconsin had this system, as he had joked about the fact that in ten years hence you could probably get it surplus. The "telescope" was a part of the ground support system for checkout of the Titan Missile Guidance System. It consisted of an equatorial mounting with servo drive and a four inch refractor. This was interfaced to the gimbal system through a rack panel containing servo amplifiers and control circuitry for a closed loop servo system. When the star tracker, mounted on the top of the gimbal system and used to update the gyros in the inertial system, locked on to a navigation star, the equatorial mount was slaved to the same position and you could look in the telescope and verify that the proper star had been acquired. Code and Houck thought that this was pretty neat and decided to instrument the equatorial mount and put a simple photometer at the eyepiece of the refractor.

We now arrive at the origin of the automatic telescope. As the hardware for OAO evolved it seemed desirable to get some experience with remote operation of a telescope. John McNall was by then the Chief Engineer and had among many other things developed the Ground Support System built around a small minicomputer (DEC PDP/8). So it was decided to set up the equatorial mounting from the Titan Missile Guidance Alignment Set at the Pine Bluff Observatory and control the positioning of the telescope with the minicomputer. This first APT was probably put into operation about 1965. It was for the most part unsuccessful. The gears on the mount were poor and the backlash very bad. It was only the large field of view that allowed it to work at all. Reworking the gearing was considered, but McNall thought it would be best to dispense with the closed loop servo system and close the loop digitally (since the minicomputer was available), and this was the basis of the system that finally evolved and did work well.

To go from an analogue servo system to a digital control system, the syncros were removed and replaced with optical shaft angle encoders and stepper motors. Terry Miedaner, a PDP/8 minicomputer programming expert, developed the code that had to fit in a very small amount of memory (only 4K of RAM). When they started operating the second system they encountered all the previous problems plus the reluctance of the stepper motors to start up.

It was at this point that they acquired their respect for digital control. Rather than going back to the shop they managed to program around all the problems. They ramped up the stepping motors, encountered catastrophic resonances, and found that the shaft encoders had a tendency to frequently misread (throw a high order bit). The resonances were avoided by digital filtering, while logic was introduced such that if an encoder reading differed from the last reading the slew would continue in the same direction until there were enough readings to vote on where they really were. The bad backlash was overcome by always coming from the same direction with enough preload to set firmly on the gear teeth. Thus ultimately McNall and Miedaner were able to program around all the mechanical problems that had been inherited from surplus gear.

The telescope housing was made from a building contractor's shed. The roof was put on rails and the side walls were dropped. This type of structure was first used when the Henyey-Greenstein Wide Angle Camera was placed at the Pine Bluff Observatory and the astronomers got tired of freezing out in the open field. When the roof for the APT system was brought under computer control it became a thing to reckon with. If you were out there when the roof motor started you had to duck in a hurry or be decapitated!

During operation as an extinction telescope (which was routinely the case), the observing sequence was preset, *i.e.*, the sequence of observations had to be in the order of the star positions on punched paper tape, and any deviations from this program had to be entered by hand. After the twilight sensor indicated that it was dark enough and (if) the rain sensor said "OK", the roof rolled back and the telescope set on the first object on the paper tape. If a star was present and the observed magnitude was within a predetermined range it was assumed that this was the correct star and it was centered first in a 10 arcminute aperture then in the 2 arcminute aperture and then observed with the 10 arcminute aperture. If no star was present, a raster scan was executed, and if that failed to acquire the star (or failed to acquire the star in the right magnitude range) the instrument went on to the next star on the list. After three successive failures it assumed that it was cloudy and closed up for an hour. Multiple observations of extinction stars required multiple entries in the input paper tape. During the night the program would continually calculate extinction coefficients and update them as data accumulated.

I feel that the Wisconsin APT was an outstanding pioneering project. It was capable of routinely finding, centering, and measuring a large number of (bright) stars across the entire sky in a totally automatic fashion with its permanently mounted photometer. This capability was not to be repeated in another small ground-based system for almost two decades, when Louis Boyd's system began routine operation in November 1983. That the Wisconsin APT was able to operate with such limited memory (only 4K of RAM) remains a marvel to this day! Although perfectly capable of measuring the brighter variable stars, the system was used to measure nonvariable stars so that nightly atmospheric extinctions could be determined. It might be recalled that Joel Stebbins was absolutely insistent on good nightly extinction measurements, and the tradition was carried on after his departure. The biggest contribution of the first APT, of course, was the ground work it laid for all the space telescopes to follow.

III. THE SKILLMAN APT

The beginnings of the David Skillman APT can be traced to his elementary school days when he built a 6 inch Newtonian reflector on a German mount. Although other projects and an interest in physics took the lead for many years, Comet Kohoutek brought the 6 inch out of the attic and renewed Skillman's interest in astronomy. Visual observing quickly paled, and an astro-camera using a portrait lens revealed mainly that such lenses are purposely built to blur the images towards the edges. About this time, Skillman met Anthony Mallama. Mallama, from northeastern Ohio, had learned photometry from the veteran amateur photometrist, Larry Lovell, and had majored in astronomy at Vanderbilt University where photometry is a specialty. A neglected photometer was found under a sink in a Goddard Space Flight Center darkroom, and an unused 12 inch telescope was borrowed from the Goddard Astronomy Club. With a photometer, telescope, and unbounded enthusiasm, Skillman was introduced to a program of serious scientific observation.

Skillman, who programmed computers for NASA by day, spent many a long cold night going back and forth between variable, comparison, and sky as he observed the eclipses of short-period binaries for times of minima. It occurred to him that a computer wouldn't mind the cold, and would have the patience to do a really good job. Thus the Skillman APT project was born. However, unlike projects while a school-aged boy, there was now a reasonable amount of money and time. Furthermore, working with NASA has instilled the value of careful planning.

The NASA approach to satellite control used a small computer on the satellite that communicated with a larger computer on the ground. Skillman similarly reasoned that he could put a small computer out with the telescope for low-level tasks in machine language, and a more powerful computer in his nearby house that would run in a high level language. He also reasoned that by specializing the system to do the simplest possible subset of tasks, success would be more likely. The simplest subset was to go back and forth between nearby variable and comparison stars (and sky background) and centre and measure these. While automating only these functions would require manual start-up, and would restrict observations to short-period variable stars (of which there are a great many in need of photometric observing), it would have the considerable merit of avoiding the difficult problem of automatically slewing to many different stars spread about the sky, and would simplify data gathering and subsequent analysis.

With these thoughts in mind, Skillman designed a computer-controlled telescope to do the specific job of automated photometry of short-period variable stars. A set of 12 1/2 inch optics was obtained from Coulter Optics, the childhood 6 inch mount was cannibalized for a polar axis, and a fork was welded up from 1/8th inch sheet steel. A third mirror was added to the Cassegrain optics to deflect the light out to the side so that the photometer could be mounted very close to the intersection of the RA and Dec axes instead of the more conventional position behind the mirror. This allowed a shorter, stiffer fork as well as good placement of the photometer.

A 6 inch diameter Byers worm gear was also salvaged from the earlier 6 inch telescope for the RA drive on the new telescope, and a second, identical gear obtained from Byers for the Dec drive. Used stepper motors (48 steps/revolution) drove a 100:1 reduction worm gear which in turn drove the 189:1 Byers gears. RA and Dec drives were identical and had a step size of 1.4 arcsecond per step, and a stellar tracking rate of about 10 steps per second — fast enough not to shake the telescope detectably.

The simplest photometer used on the system was built from brass tubing, left-over 35mm film cans, and a 1P21 bought from a local radio store. A 30 arcsecond diaphragm was used, and the manually changable *UBV* filters usually were left on *V* for automatic operation. A DC amplifier and analogue/digital converter provided the digital photometric input to the computer.

The entire system (not counting labour) cost about $4000 (1980), with the largest expense being $2000 for the Apple computer. The software for the KIM computer (out at the telescope) was written and assembled by hand, as an editor/assembler was not available. When the system was fired up in 1979, it worked the first time and within a week or two was routinely gathering data on times of minima of eclipsing binary stars.

An article on Skillman's APT in *Sky and Telescope* (Skillman, 1981) was widely read by many, including astronomers at Steward Observatory who later hired Skillman to program the microcomputer for their 90 inch telescope on Kitt Peak. Later, as was the case for several of those involved in the Wisconsin APT, Skillman was asked to work on the Space Telescope, where he is today.

Throughout this entire time, the Skillman APT continued to build up an impressive observing record. Many of the observations of eclipsing binary stars were published in the *Journal of the American Association of Variable Star Observers.* Short-period RS CVn binaries were also observed, and most recently V 471 Tau was observed for hundreds of hours (a paper is in preparation).

The major change to the system since its initial operation is the replacement of the original simple photometer with a "smart" photometer that contains its own microprocessor (Z-80). The photometer can execute filter sequences, change diaphragm sizes, and store millisecond-spaced readings at high speed for later low-speed playback.

Skillman was able, working on his own on a very low budget, to use two of the early microcomputers to automate most of the observational functions — allowing operation (once started up) while he slept. His system has been gathering valuable photometric data on variable stars for five years, and is continuing to do so every clear night. This fine record of accomplishment speaks for itself.

IV. THE BOYD APT

Although born in Indiana, Louis J. Boyd moved to Missouri at a young age, and the dark countryside nights kindled an interest in astronomy in both Louis and his father. Louis loved to build gadgets and it was not surprising that he majored in engineering (electrical) at the Missouri School of Mines and Engineering. A tour in the Army ended at a fort in southern Arizona and, on returning to civilian life Boyd moved to Phoenix and worked for the Mountain Bell telephone company. It might be noted that much of the telephone equipment he worked on was fully automatic and placed on remote mountain tops, where it operated unmanned for weeks or months between visits by telephone company personnel.

While scrounging electronic parts at a surplus outlet (for a radio telescope project that was later given up in favour of photometry), Boyd met another scrounger, Richard Lines. Richard and Helen Lines were avid amateur astronomers, with a 16 inch telescope at Mayer — a dark site location between Phoenix and Flagstaff. As might be guessed, Richard Lines was scrounging for parts for a photometer, and enlisted Boyd to help him on the project. With the photometer working, Richard making the observations at the telescope, and Helen giving directions and writing down the data, photometry at the Lines Observatory quickly became the highly-ordered and precise operation it has been noted for ever since. One night, Boyd, who was visiting and watching the highly-systematic and repetitious operation, suggested that it could all be computerized. Helen Lines, with a cryptic suggestion, said, "Why don't you stop talking about it and go do it?" This was the beginning of the Boyd APT.

It is interesting to note that Boyd had no background in doing manual photometry, nor did he acquire any during or after his project. He remains the only photometrist I know of who has only made automatic observations. Nor was Boyd aware of the earlier Wisconsin or Skillman efforts. His approach was (and remains) one of applying basic engineering skills to a carefully-defined and restricted problem to come up with straightforward and simple solutions. Only those capabilities required to accomplish the direct job at hand were included — no frills were allowed.

Two years into the project, the capability of doing photometry on a single groups of stars (variable, comparison, check, and sky) with initial manual acquisition was achieved. It might at first seem that going from this to doing a large number of different stars across the sky would not be difficult, but this was not at all the case. Another two years elapsed before the long-sought goal of completely automatic photometry on a large number of different stars was achieved. With Boyd's approach, a list of variable stars (along with comparison stars, sky positions, etc.) to be observed throughout the year is entered just once. The system itself then decides, as each night progresses, just what variable star it should observe next. The only human intervention in this process is one of occasionally pulling off diskettes full of data, doing periodic cleaning of mirrors, and any other maintenance. It is this total and complete automation of the entire process that is the most important of Boyd's several contributions to APT's. This system was first described by Boyd, Genet & Hall (1984).

It seems pertinent to note that Boyd's original APT has been a work-horse — not just a development item to be discarded when improved versions came along. The record of observations and publications from this one automatic telescope located in the city of Phoenix is nothing short of phenomenal. There are no plans, for the foreseeable future, to shut this system down. Thus every clear night (200 or so per year) this system will continue to observe all night long without pause. Rather than stop this system to make improvements, these were channeled into the second generation system described next.

V. THE FAIRBORN OBSERVATORY APT

The Fairborn Observatory APT was a group effort, and it was a direct descendant of the Boyd APT. In May, 1981, just prior to a run at Kitt Peak National Observatory, I visited a group of Phoenix-area photometrists who had gathered at Jeff Hopkins' observatory for the evening. Among them was Louis Boyd, and we discussed his APT project. This visit initiated a strong interest on my part in Boyd's project and it was followed in considerable detail via letters and phone calls.

In November, 1983, I arranged to be present in Phoenix for the first full night of automatic operation of Boyd's APT. I was very impressed, and while flying back to Ohio, I decided to assemble a system with the same capabilities as Boyd's but in a manner that would allow full documentation and potential duplication by others if desired.

Boyd quickly agreed to design a second generation control system that would have the same capabilities, but would be greatly simplified. It was based on the peripheral technology PT-69 single-board microcomputer — a 4 x 6 inch computer costing $300 (1983), not including power supply, disk drives, and cabinet. All the other digital electronics were placed on a single wire-wrapped board, also 4 x 6 inches in size. Two simple high-speed stepper drivers (one for each axis) were hand-wired on small perf boards by my son Michael.

About two-thirds of the software from the (first generation) Boyd APT was usable without any changes whatsoever, even though the control electronics had been completely changed. (An obvious advantage of modular software.) New software was developed by Boyd, Lloyd Slonaker (a computer science student at Wright State University), and myself.

For some years I had been trying to convince Frank Melsheimer, President of DFM Engineering, that besides his medium-to-large-sized professional telescopes (18 to 94 inches), he should design and produce a smaller-sized low-cost telescope for use by small colleges and advanced amateurs. The DFM Engineering mounts, with their zero-backlash friction drives and extremely rigid structure, seemed ideal for automation. When the Massachusetts Institute of Technology ordered a right ascension only CCD camera platform for their gamma ray burster project, the basis for a low-cost mount using aluminum castings was laid. Melsheimer kindly went on to build a second mount and added to it a fork and declination friction drive system. This system — the prototype mount of the DFM Engineering Small Telescope System — was kindly donated to the Fairborn Observatory for use on the second generation APT. This mount and the entire DFM Engineering Small Telescope System are described in some detail by Melsheimer & Genet (1984).

A compact optical system (low inertia) seemed appropriate, as did a completely closed tube to withstand the rigours of the Ohio climate. John Diebold, President of Meade Instruments, kindly donated the 10 inch Schmidt-Cassegrain optics used on the APT.

The photometer used on the APT was the rugged and trouble-free Optec SSP-3 — kindly donated to the project by Gerald Persha, President of Optec. This photometer has a $BVRI$ capability, although the system initially operated in V only. This photometer was later replaced with the SSP-3a which is equipped with a 5-position filter slide controlled by a small stepper motor.

Lloyd Slonaker helped me integrate the entire system, and the first fully automatic operation on a program of variable stars was achieved on the night of 19/20 September 1984 — 10 months and two weeks after the Boyd APT saw its first all-night operation. This system was later disassembled and moved to Arizona. It has been reassembled and installed on Mt. Hopkins — just down the ridge from the Multiple Mirror Telescope (MMT).

This APT has been fully documented in the book *Microcomputer Control of Telescopes* (Trueblood & Genet, 1985), where three chapters describe the equipment and software, and appendices contain algorithm details and software listings. (These include listings of observational reduction and various support programs.) If one wished to duplicate this system, the computer, telescope, and software are all "off-the-shelf" items, while the control electronics and stepper drives could be reproduced by a reasonably experienced electronics technician or engineer.

VI. CONCLUSIONS

As this is billed as a historical review, it seems only proper to end on a broad historical note. There was a revolution in photometry in the mid-1940's, initiated by Gerald E. Kron when he introduced the use of the 1P21 photomultiplier, DC amplifier, and strip chart recorder as an integrated photometry system. While there were precursors to this, such as the use of amplifiers by A.E. Whitford, it was Kron's "system", first used at Lick Observatory immediately after the war, that revolutionized photoelectric photometry. Prior to this, photometry was done at only a few observatories. The equipment was difficult to work with, prone to failure, and expensive. Afterwards, it was easy to use, reliable, and relatively cheap. Harold L. Johnson and others did much to publicize and apply this new capability to good advantage, and the golden age of astronomical photometry was entered. Whether or not the computerization of photometry will eventually be seen as another revolution in photometry must await the more seasoned judgement of those in the future, but the parallels are interesting.

The Wisconsin APT is startling for its earliness in time. Not only was it pioneering as an APT, but it was pioneering as a digital servo system. That it was achievable with so little computer memory (only 4K of RAM) is both a marvel and an indication of the difficulties imposed by the technological limitations of the time. That the system was not used for variable star research is probably more a reflection of the interest of its creators in space telescopes than on the capabilities of the system (it would have been limited to very bright variable stars with a 10 arcminute diaphragm).

The long gap between the Wisconsin APT (fully operational in 1967) and the Skillman APT (not operational until 1979) is indicative of the time taken for microcomputers to become reasonably inexpensive and available. It is also a result of the bias inherent in this paper towards small telescopes and systems that worked for a period of time. There were larger systems — some which worked somewhat and some that did not. Be that as it may, microcomputers were the great equalizer — being so low in cost that amateurs, automating small photoelectric telescopes for enjoyment, could dominate the next phase of APT development.

The Skillman APT became operational in 1979 — as early as one might reasonably expect for microcomputers. The need for hand coding and assembling machine language for the KIM microcomputer is indicative of some of the difficulties in these early days of microcomputers. By employing manual startup and limiting the system to continuous monitoring of short-period variable stars, Skillman was able to achieve a solid success.

The Boyd APT, like the Wisconsin APT, was fully automatic, not requiring manual startup, and thus like the Wisconsin APT was capable of photometry of many different stars spread about the sky. With a large memory (64K of RAM) and other benefits of advanced technology such as disk drives and high-level languages usable in real-time control, the Boyd APT was able to go beyond a fixed observational sequence, or one limited to one small group of stars, to a flexible observing program on many groups of stars. Observational requests from research astronomers were placed on the observing program, and observational results were returned in a routine manner — a completely computerized process from beginning to end.

The Fairborn Observatory APT really had no new capabilities at all beyond the Boyd APT – in fact, it used much the same software. Its uniqueness lay in its use of an off-the-shelf mount, optics, and photometer, the simplicity of its electronics, and the completeness with which it was documented. It signals, perhaps, the end of the early APT era, and, from its home at the APT Service on Mt. Hopkins, points the way towards large scale automatic observations being made from remote mountain tops for observatories around the world.

REFERENCES

Boyd, L.J., Genet, R.M., & Hall, D.S. (1984). *IAPPP Comm.,* **15**, 20.
McNall, J.F., Miedaner, T.L., & Code, A.C. (1968) *A. J.*, **73**, 756.
Melsheimer, D.F. & Genet, R.M. (1984). *IAPPP Comm.,* **15**, 33.
Skillman, D.R. (1981). *Sky and Telescope*, **61**, 71.
Trueblood, M. & Genet, R.M. (1985). *Microcomputer Control of Telescopes.* Richmond: Willmann-Bell Inc.

NEW DIRECTIONS IN THE STUDY OF VARIABLE STARS WITH SMALL TELESCOPES

S.M. Rucinski
David Dunlap Observatory
Department of Astronomy
University of Toronto
Toronto, Ontario
Canada M5S 1A1

I. INTRODUCTION

The present assessment of the new directions in small telescope observations will be dominated by possible trends in the instrumental developments, which are expected to improve both the accuracy and the efficiency of observations. As with any such predictions, one must rely solely on the present state of affairs; the real future might in fact look entirely different.

In what follows, we will make a distinction between observations by *amateurs* and in *small semi-professional institutions*, like colleges and small universities. In the former case, the time and resources might be expected to be more limited but the main distinguishing factor for our purposes is the *interest* in performing observations which — hopefully — should rather quickly give interesting and useful results. In the latter case, one might expect more emphasis on the *educational* aspect. The ensuing data reductions may — or perhaps even should — be more complicated, so as to expose students to the intricacies of astronomical data handling, reduction and interpretation. An obvious tendency might then be a dominance of photometry in amateur observations and spectroscopy in small college observatories.

As to the location of the actual observations, we shall assume that it is not an astronomically advantageous one. We therefore exclude mountain-peak observatories and address those observers who want to obtain good quality data in an (astronomically) poor climate, *i.e.*, with frequent occurrences of cirrus clouds, with variable and large extinction, and poor seeing. Useful hints and suggestions on how to perform photometric observations in such conditions are contained in the article by Fernie (1983). As he advises, the best solution is to use two independent telescopes, one for observing program stars, the other to monitor reference (comparison and check) stars. A similar system of two telescopes, but mounted on one pier, has recently started to produce good-quality data in Scotland (Bell & Hilditch 1984). The two-telescope solution (Figure 1) may be a better solution than a double-channel photometer on a single telescope. Judging by a large number of such double-channel single-telescope systems in existence, and their relatively modest output, such systems would seem to be considerably more difficult to operate and — more importantly — to calibrate. This is becaause of the non-uniform photometric properties of the field of view in most telescopes (the channels may have to be re-calibrated in sensitivity for each separation of the photometer heads). Some of them, however, seem to give amazingly good results (*cf. e.g.*, Grauer & Bond 1981).

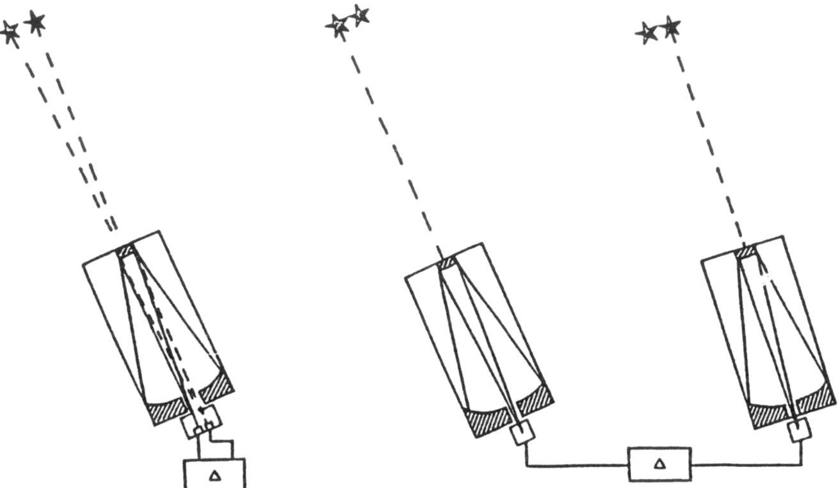

Figure 1. Two designs for a two-star photometer system.

In effect, spectro*scopic* observations are naturally differential in nature, since comparison of one part of the spectrum is made with other parts. As such, the spectroscopic observations do not require that the weather be especially good (this is in contrast with spectro*photometric* observations which can be done only in the best astronomical climate). The photographic plate has a long and honorable history as the detector for such observations. However, its capabilities in terms of attainable accuracy seem to be close to a limit. The signal-to-noise ratio for photographic plates cannot exceed a figure higher than about 50 but — more typically — it is about 30 (Smith & Hoag 1979); also, their photometric response is non-linear and dependent on the exposure time. These deficiencies are well known. In spectroscopy, they have led to the development of *scanners* which — in their differential versions — employ either a number of detectors which simultaneously see different parts of the spectrum, or single detectors with the spectrum quickly moved in front of them by, say, wobbling of the spectrograph grating (Figure 2).

Both photometric and spectroscopic observations will benefit enormously from the development of *panoramic detectors*. In effect, a modern panoramic detector acts as a photographic plate but with much better — and controllable — signal-to-noise characteristics and with a linear response over a large dynamic range. The present properties of these detectors will be briefly summarized in the next section.

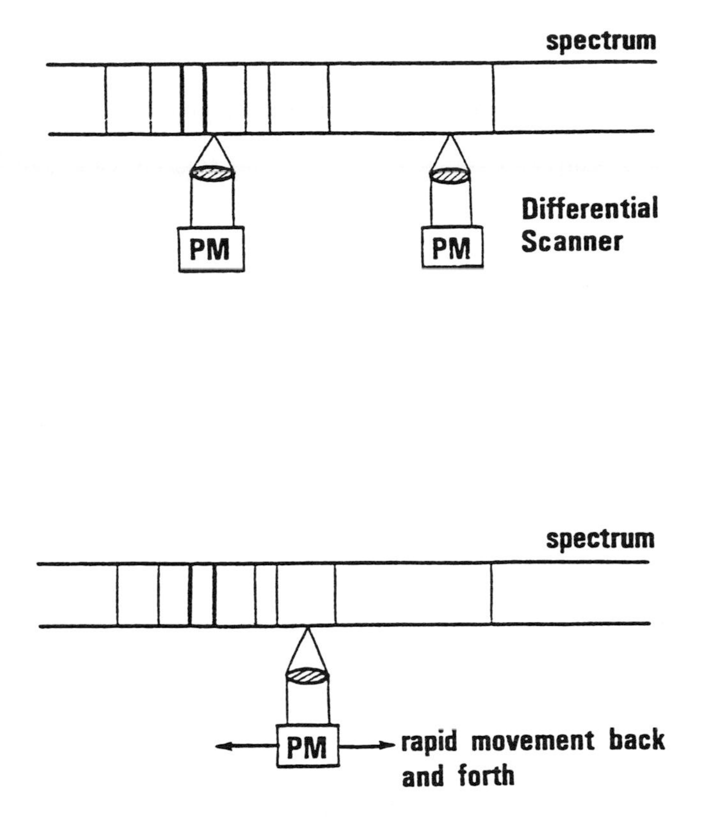

Figure 2. Two methods for recording a spectrum non-photographically.

II. PANORAMIC DETECTORS

The term "panoramic detector" will be used here to mean a rectangular array of many separate detectors which are capable of recording images of the sky or spectrum electronically. At present, the semi-conductor, charge-coupled devices (CCD's) can serve as a good example, but so much effort is now being spent on the development of other new detectors that even the modern CCD's may soon become obsolete.

A state-of-the-art CCD array typically consists of 1000 $\times$ 1000 adjacent, very small (say, 10 μm) picture elements (pixels) each capable of recording signal in a dynamic range at least 10^5:1. The electrons generated within a given pixel (with the quantum efficiency now approaching 80 per cent: 8 electrons per 10 incident photons) are stored there for selectable exposure times which can range from a fraction of a second to many hours. The read-out is into a computer memory, which contains a one-to-one mapping of the original image. The present CCD detectors are limited in noise performance by the electronic read-out noise, which is (at best) at the level of 10 electrons per read-out.

The sensitivity of the CCD detectors is really impressive. Converting the above numbers to more familiar units, one obtains that using a 60 cm (24 inch) telescope, a photometric accuracy of 1 per cent (0.01 mag.) in a photometric observation of a $V = 15$ magnitude star can be achieved in just 15 seconds! Accuracy of 0.3 per cent (0.003 mag.) requires a 2.1 minute exposure.

Modern large CCD's are now quite expensive (up to 50 K$), require expensive controllers (another 50 K$) but — judging by generally rapidly falling prices of digital hardware — these numbers should improve on time scales of a few years. Even now, however, small CCD's cost much less than the large-size versions, but it would be hard to establish any simple relation between the sizes and prices from the present strongly fluctuating prices. (It would not be surprising if the prices depend not on the surface area but say, as a cube of the side dimension.) On the other hand, electronically inclined amateurs may contemplate building their own controllers, thus reducing costs even further. In fact, descriptions of inexpensive CCD systems built by amateurs have already started to appear (*e.g.*, Bull 1985).

The main problem with CCD's is that they require cooling. When uncooled, the detectors generate an unacceptably high dark current and read-out noise. One might hope that the cooling requirement will improve with time, but certainly not by a great deal, since the dark current in semi-conductors cannot be eliminated at normal ambient temperatures. Thermo-electric coolers are the simplest solution, but then the main "higher-order" problem is to keep the same temperature across the whole area of the array. The reason is that even small temperature gradients produce large differences in sensitivity and relative spectral response, which are difficult to remove in later stages of the image reduction and analysis.

III. OBSERVATIONS OF VARIABLE STARS WITH PANORAMIC DETECTORS

One can think of small-telescope observations with panoramic detectors as a return to the old-fashioned photographic observations: the variable and comparison stars are observed simultaneously; their images are first recorded and then analyzed later on, to derive the differential magnitudes. But there are important differences in the equipment:

a) There must exist enough memory storage for the image. If we use a 1K × 1K array, the storage requirement is at least 1Mbyte (if each pixel is stored as one byte only).

b) The display should be large enough to show the picture.

c) The image should be stored on some convenient medium, preferably in a single writing process, without further possibilities of over-writing the record.

At present, micro-computers which fulfil the above requirements are quite expensive but should become less so on a short time scale. The magnetic storage media, although reasonably reliable by present standards, will soon be replaced by even better *laser disks*. The video disks of this type should be able to store of the order of 1Gbyte (or 1000 images of 1Mbyte each!) and should be particularly useful as a "write-once" medium for long-term storage.

One can think that in the not-too-distant future, observations of variable stars will look like this: A panoramic detector, coupled to a computer and a storage device, will permit taking pictures of a stellar field containing a variable and a number of comparison stars. These images will obviously contain information on the relative brightnesses of all stars in the field, irrespective of the changes in the atmospheric transparency and in the background level. Later on, one will analyze these pictures by measuring intensities of stellar images, by interpolating and removing the background, etc. Very little observing time will be wasted because images can be taken in succession, with little time lost in between. The subsequent analysis can be done many times, by many people, using different extraction algorithms. Thus, the role of an observer would be only to take pictures; the analysis would be done later in the best and most optimal way. One might hope that this will avoid the dependence of the quality of observation on the observer — a situation which is quite common now. Indeed, this quality now depends very much on *how* the photoelectric observations are done: how often the comparison and background measurements are done, what is their succession, what is the duty-cycle of the observer/telescope combination, etc. This unfortunate circumstance has led to the situation that some observers are considered "good", some "bad", only because the methods by which they do their observations differ. The panoramic detectors will enable everybody to analyze the same observations, in a similar way as it used to be possible with photographic photometric observations.

IV. SPECTROSCOPY WITH PANORAMIC DETECTORS

Spectroscopy and, in particular, high-resolution spectroscopy, is very attractive for observatories located in poor climates. A single spectrum or even a single spectral line can contain a very large amount of astrophysical information. However, until very recently, high-resolution spectroscopy was restricted to observatories having large telescopes. Now, partly due to the development of very sensitive panoramic detectors, this type of observation can be contemplated by small institutions with modest financial resources.

The need for routine spectroscopic observations is now felt particularly strongly in stellar astronomy. Because of the inflexible scheduling of large telescopes and the heavy demand on observing time for extra-galactic objects, it is now difficult to monitor stars (especially variable stars) spectroscopially in the way it is possible to do photometrically (to a large extent thanks to the continuing efforts of the AAVSO and IAPPP members). Many interesting photometric observations have no spectroscopic support, and are therefore more difficult to interpret. In fact, the whole discipline of high-resolution stellar spectroscopy has fallen in some disfavour among astronomers; the sorry state of this discipline is described in the review paper by Wolff (1983).

Two types of spectral instruments which utilize the high efficiency of panoramic detectors are within the reach of small astronomical institutions:

a) the slitless spectrograph,

b) the echelle spectrograph.

As pointed out by Furenlid (1984), the simplest astronomical spectrograph of high efficiency could consist of just one concave diffraction grating without any other optical elements (Figure 3). The light is perfectly collimated by the remoteness of the stars, and then focussed on the panoramic detector by the grating. According to his estimates of one particular combination of the grating parameters, "at an aperture having an equivalent diameter of 7.3 inches, the instrument has approximately the same limiting magnitude as a conventional 60 inch telescope and spectrograph of the same dispersion" (here the dispersion is 14.3 Å/mm). However, we must remember that there exists a weak point of the slitless spectrograph: the dependence of its resolution on the seeing.

A much more complicated instrument is the *echelle spectrograph.* As it is usually mounted at the Cassegrain focus, where flexure problems are the most common, it requires special attention to the mechanical stability and rigidity. It is also relatively large and considerably more difficult to build. However, it is currently the best answer to the need for obtaining high-resolution spectra at small and moderate size observatories. The high dispersion is achieved in this type of spectrograph not by increasing the sizes of the grating and the remaining optical elements (as is necessary in conventional systems) but rather by combining the dispersive powers of two diffraction gratings. One of the gratings, the "echelle", works in the high order of interference (typically 10 to 100); the other helps in separating the otherwise overlapping segments of the spectrum by shifting them in the direction perpendicular to the dispersion of the echelle (Figure 4).

The echelle spectrograph has a few very convenient features. One is the ease of obtaining high-resolution spectra with a spectrograph whose elements have rather

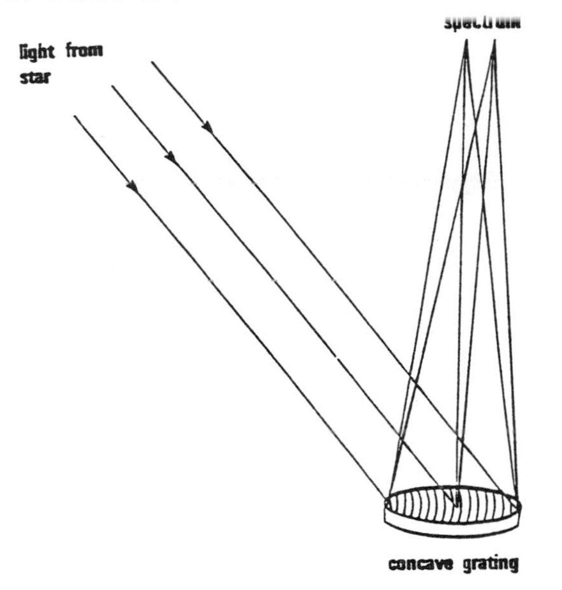

Figure 3. One single component, a concave grating, may be used to generate a focused spectrum of astronomical objects.

moderate dimensions. Thus, contrary to the standard design, the collimator and camera focal lengths do not have to be large and there is no need for particularly fast cameras. The moderate collimator focal length permits a wider entrance slit so that more light is admitted into the spectrograph even in conditions of poor seeing. Another convenient property is the format of the spectrum, consisting of parallel segments which can be relatively evenly distributed over the sensitive area of a panoramic detector (*cf.* the nice echelle spectra published by Hearnshaw (1978) and Latham (1978)).

Most of the echelle spectrographs built to date have had parameters chosen to match the performance of coudé spectrographs with typical dispersions of, say, 2 to 4 Å/mm. This aim is not always easy to achieve, so quite a number of such units do not find as much use as they should due to their mechanical problems. But the echelle spectrographs could also be built for lower dispersions; then they are much easier to build and maintain. In addition, the development of fibre-optic cables should solve the mechanical problems with high-resolution spectroscopy. A spectrograph can now be entirely disconnected mechanically from its feeding telescope: it can rest on a solidly-built support somewhere in the dome or even in some other room, and the light can be sent to it through the fibre-optic cables.

Collections of useful formulae and hints for construction of an echelle spectrograph can be found in papers by Schroeder (1970) and Schroeder & Anderson (1971).

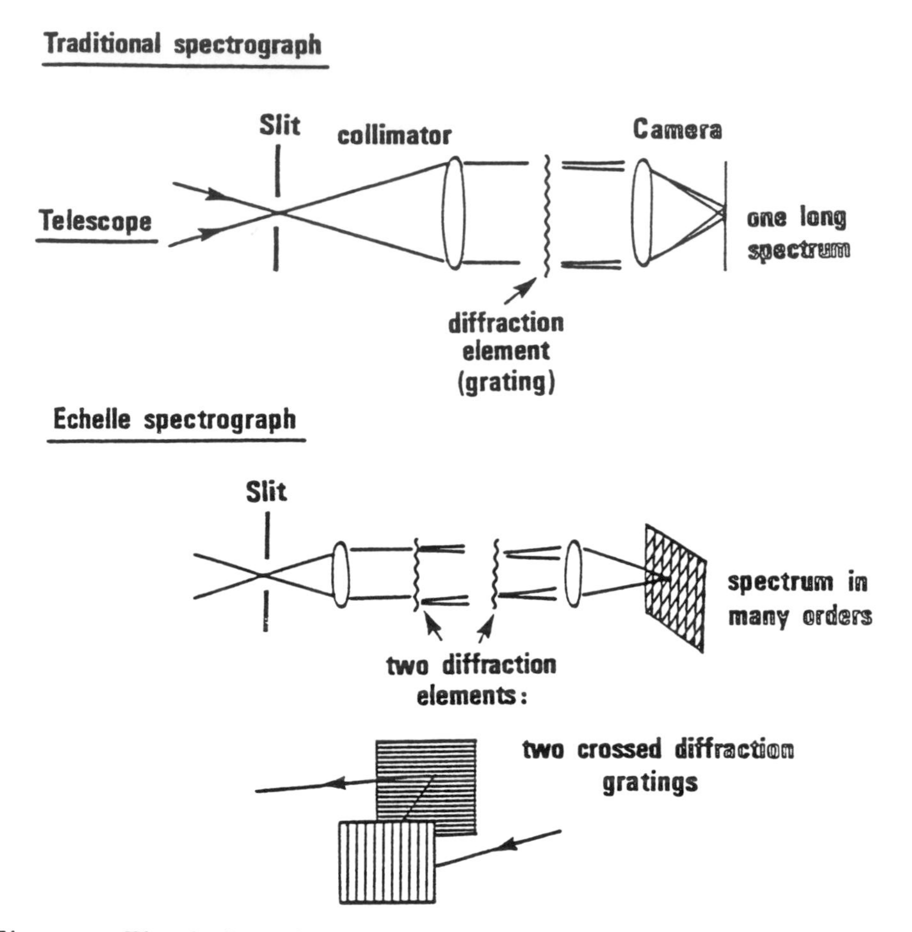

Figure 4. The design of the traditional and echelle spectrographs.

V. LUNAR OCCULTATIONS

Superficially, observations of lunar occultations do not seem to relate directly to research on variable stars. However, there are important links between these disciplines, and the most direct connection is through the use of exactly the same photoelectric photometer as that used to observe variable stars. There do exist special requirements on the data-acquisition system, which must be able to handle data at millisecond intervals. It is expected, however, that the interest of actively observing amateurs will evolve in the direction of more advanced electronics, so that the lunar occultation technique may eventually become quite popular among them.

Two excellent articles on the subject, by Blow and by Chen, appeared in the same book (ed. Genet 1983). Both are addressed to observers having small telescopes, so only a few remarks are in order here.

About 11 per cent of the sky is available for such observations, which are of a very special nature: the Moon, by occulting a star, produces a diffraction pattern of the star's light which moves rapidly on the Earth. This shadow pattern, whose width is about 50 - 100 meters, contains information on the angular size of the star or on its possible duplicity in the "difficult" range of angular sizes between roughly 0.005" and 0.5". Motion of this pattern is really very rapid, with the speed corresponding to the orbital velocity of the Moon, *i.e.*, about 0.5 meter per millisecond. Thus, everything takes place in a fraction of a second, and must be recorded that fast by setting the data acquisition rates of the recording equipment appropriately.

The occultation events, which are in fact half-ready measurements of stellar diameters and binary separations, are performed for us at irregular intervals of time, and many of them are missed. However, it is really a big waste not to use these opportunities, because we must remember: the occultation method is the only one which gives an angular resolution greatly exceeding that of the diffraction resolution of the telescope!

The help of amateur observers is needed for many reasons. One reason is the nocturnal visibility and the weather, which do not always cooperate. The other reason is related to the large increase in accuracy when the same occultation event is observed from different locations. This is the result of the rugged shape of the Moon's limb at angular scales comparable to the wiggles in the diffraction pattern. Small hills and even large boulders with sizes between roughly 10 and 100 meters are particularly annoying, because their effect appears as an unexpected change (stretching or compressing) of the time scale of the diffraction record. The best solution is to observe occultations with at least two telescopes separated by a few tens of meters in the direction *perpendicular* to the direction of the shadow motion. One should also keep in mind that the data-gathering efficiency can be strongly enhanced by observing the same event with the same telescope but simultaneously in more than one *wavelength*. The reason for such an approach is simple: the diffraction patterns in different wavelengths have exactly the same shape but their time scales are re-scaled (are stretched or compressed) by the factor proportional to the square root of the wavelength.

The measurements of stellar diameters and binary separations using the lunar-occultation technique have not yet reached the stage when the time variability of diameters or binary separations could be routinely studied. However, the stellar diameters are most easily determined for late-type (K and M) giant and supergiant stars, which almost exclusively belong to the category of pulsating variables. Hence, one might expect that the method will enhance and supplement in future the current methods of studying these stars, which are based solely on the combination of photometric and spectroscopic data.

VI. THE FOURIER-TRANSFORM SPECTROGRAPH

This instrument does not look like a spectrograph and its immediate output does not look like a spectrum either. Yet, it may be the ultimate spectral instrument for astronomical applications, surpassing in performance the best of traditional spectrographs. Indeed, those few Fourier-transform spectrographs (FTS) which are in operation now produce very high-resolution spectra of unbelievable integrity over very large ranges of wavelengths (Ridgeway & Brault 1984).

The scheme of an FTS is very simple (Figure 5). By moving one of the flat mirrors in a controllable way, one can introduce an optical path difference x to the two light beams, which interfere on the beam-splitter. For a monochromatic wave with the wave number $k = 1/\lambda$ this results in a phase difference $2\pi kx$. When such a wave of intensity I_{in} enters the instrument, the detector registers an intensity:

$$I_{out} = I_{in}[1 + \cos(2\pi kx)]/2$$

Thus, the signal is now modulated according to the relative positions of mirrors, which enters via the optical path difference x.

The real light consists of a mixture of different waves forming a spectrum $I_{in;k}$, the actual object of our interest. Because of this contribution of different waves with different wave numbers k, one sees at the output the signal:

$$I_{out;x} = A/2 \int_0^\infty I_{in;k} \cos(2\pi kx) dk + B$$

The constant B is not interesting because it represents the part of the signal which does not change with the movement of the mirrors. The constant A, on the other hand, represents various light losses in the instrument; obviously, it should be kept as close to unity as possible.

The integral expression above is one of the forms of the Fourier transform, and this is the reason for the name of the instrument. To recover the spectrum $I_{in;k}$ from $I_{out;k}$ one must perform the *inverse* Fourier transform and this can now be easily done on practically any computer.

The main advantage of an FTS is that the whole spectrum is observed at each moment, *i.e.*, at each position of mirrors of a given x. The instrument is therefore a truly multiplexing one, as all frequencies in the light are observed simultaneously. To decode them and to recover the spectrum, one must change x and the best way to do that is by a scanning procedure which is rapid enough to avoid changes in the atmospheric transparency.

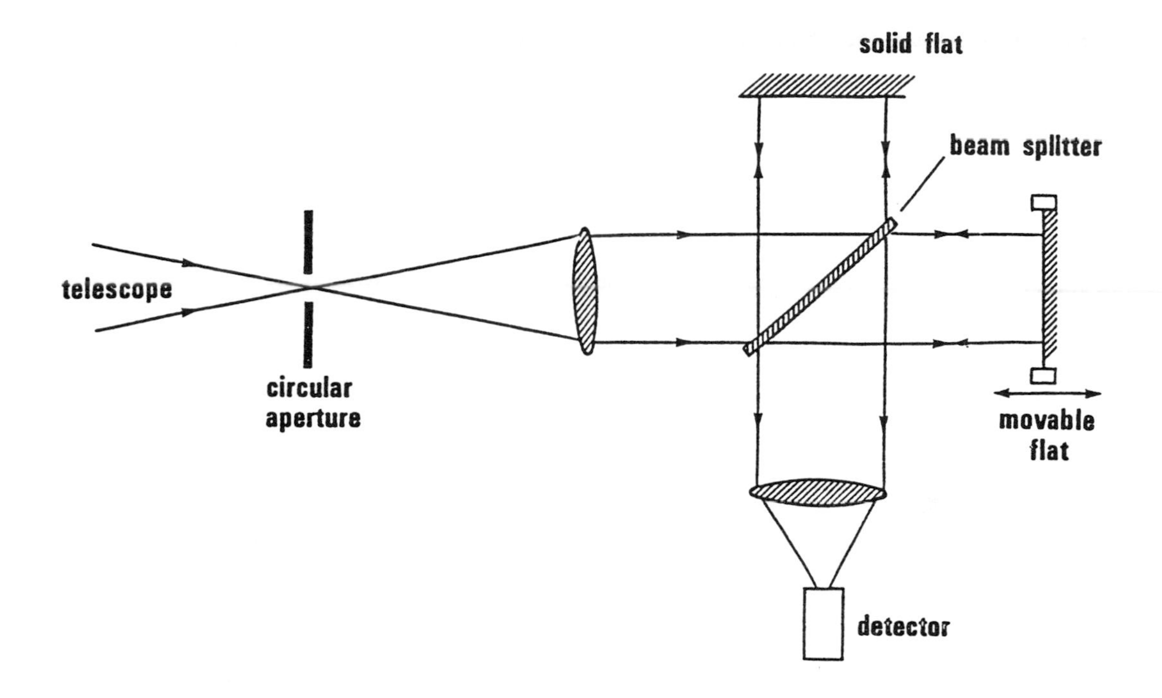

Figure 5. The design of a Fourier-transform spectrograph.

Another advantage of the FTS is that its entrance aperture may be quite large and has no relation to the instrument's resolution. Obviously, one does not want to observe too much spectrum of the sky, so a reasonable choice must be made, but even a large sky contribution can be easily eliminated by simultaneously observing through a second (sky) aperture. The large-aperture advantage increases the overall efficiency of the FTS relative to a traditional spectrograph by a very large factor, perhaps even a hundred times or so.

The lack of any dispersive element in the FTS also means that the spectra are much "cleaner", without spectral "ghosts" and any scattered light filling in deep absorption lines. And finally, the same instrument can have different resolutions depending on the way the scanning in the optical-path-difference x is performed: for bright objects, one can select a dense stepping in x which leads to a high resolution; for faint objects the resolution may have to be reduced to retain reasonable exposure times necessary to accumulate a sufficient signal.

The FTS is certainly not easy to build. It must be made to interferometric mechanical tolerances. The mirrors of the size of, say, 10 cm must freely and quickly move over comparable distances without any tilt or distortions. Their instantaneous positions must be known to μm-accuracy with the help of a laser beam. These requirements are especially demanding when the scanning is rapid to compensate for atmospheric variations. It is somewhat easier to build an FTS for the near-infrared region than for the visual region. The sampling in x may then be in larger steps; in addition, one more advantage of the FTS (called the Fellgett's advantage) can then be realized. This advantage is related to the noise characteristics of the detector: it applies when the noise is unrelated to the light level (as in the IR detectors) but disappears when the noise is determined primarily by the statistics of the incoming photons (as is the case with visual region detectors). However, even without this advantage, the FTS has a much higher efficiency than a traditional spectrograph of the same resolution and gives much better spectra.

Construction of an FTS can be contemplated by colleges which have a strong physics department support. A relatively simple system has been described by Thompson & Reed (1975). General references can be found in Ridgeway & Brault (1984) and a more technical description is given in the article by Schnopper & Thompson (1974).

VII. WHAT TO OBSERVE?

It would not be an exaggeration to say that all variable stars require further observations: deviations from strict regularities are a rule rather than an exception, so that we need high-accuracy observations which are distributed in time densely enough to pick up any variations. (One can even say that any star is variable at some level of observational accuracy!)

It would not seem appropriate to repeat here the nice review of important classes of variable stars by Mattei which is published in this same book. Rather, I will give my subjective view of two extremes: where the small-telescope observations are most needed and where they are needed now but might quite quickly become obsolete or unnecessary.

I think that the field of semiregular, long-period, and Mira variables is the most important one for small-telescope observers. Certainly, there are still pockets of interest in other areas of variable star research which are available for serious contributions, but the red pulsating variables is the area in which the help of small-telescope observers is especially needed. These variables offer an insight into the poorly understood phenomenon of the interaction of stellar pulsations with convection. Turbulent convection in stars in general defies a proper theoretical description;

its understanding becomes particularly difficult in the presence of stellar pulsations. Both phenomena have a basic importance for the physics of stellar hydrodynamics, and red pulsating stars are the best laboratories for their studies. However, these stars need very long series of evenly distributed observations of similar accuracy, and these requirements can seldom be fulfilled by professional astronomers. At present, photometric observations must suffice, but it would be extremely useful to have long series of spectroscopic observations as well.

In fact, a spectroscopic monitoring program could be profitably performed for practically any class of variable stars, including even such seemingly repeatable variables as eclipsing binaries, where mass-transfer and stellar activity phenomena are very common. Obviously, there is no point in stressing here the importance of such spectroscopic monitoring for eruptive variables like novae and supernovae, where *any* spectrum is of great value for interpretation.

Contrary to the spectroscopic monitoring programs, one can expect that the *discovery* of erupting stars can very soon be dropped from lists of profitable small-telescope projects. Very much depends on whether an observatory or even an individual would actually like to build a nova/supernova detection system. Such a system is entirely within the reach of contemporary technology and, I am sure, will be built sooner or later. One can easily imagine a telescope which jumps from one field of the sky to the other (or from galaxy to galaxy) and produces electronic pictures with its panoramic detector. These pictures are then compared automatically with those taken previously, to see it anything new has appeared in the field. Such a differential comparison is much simpler than a real "pattern recognition" which is indeed a complicated process and does not have to be done in this case. An instrument of this sort does qualify to be considered in this meeting because — to have a large field of view — it would have to be based on a small telescope. The main emphasis in its construction would be the data acquisition and the automatic comparison of images. One might suspect that a small number of such systems would be entirely sufficient to keep watch on all new erupting variables.

Acknowledgements

This contribution was supported by operating grants from the Natural Sciences and Engineering Research Council of Canada to C.T. Bolton, J.D. Fernie, S.W. Mochnacki, and J.R. Percy.

REFERENCES

Bell, S.A.& Hilditch, R.W. (1984). *M. N. R. A. S.*, **211**, 229.
Bull, C. (1985). *Sky and Telescope*, **69**, 71.
Fernie, J.D. (1983). In *Advances in Photoelectric Photometry*, Vol. 1, ed. R.C. Wolpert & R.M. Genet, p. 59. Fairborn OH: Fairborn Observatory Publ.
Furenlid, I. (1984). *Pub. A. S. P.*, **96**, 325.
Genet, R.M. (ed.) (1983). *Solar System Photometry Handbook*. Richmond VA: Willmann-Bell, Inc.
Grauer, A.D. & Bond, H.E. (1981). *Pub. A. S. P.*, **93**, 388.
Hearnshaw, J.B. (1978). *Sky and Telescope*, **56**, 6.
Latham, D.W. (1978). *Sky and Telescope*, **56**, 8.
Ridgway, S.T., & Brault, J.W. (1984). *Ann. Rev. Astr. Ap.*, **22**, 291.
Schnopper, H.W. & Thompson, R.I. (1974). In *Methods of Experimental Physics*, Vol. 12A, ed. N. Carleton, p. 491. New York: Academic Press.
Schroeder, D.J. (1970). *Pub. A. S. P.*, **82**, 1253.
Schroeder, D.J. (1971). *Pub. A. S. P.*, **83**, 438.
Smith, A.G. & Hoag, A.A. (1979). *Ann. Rev. Astr. Ap.*, **17**, 43.
Thompson, R.I. & Reed, M.A. (1975). *Pub. A. S. P.*, **87**, 929.
Wolff, S.C. (1983). *Pub. A. S. P.*, **95**, 529.

INDEX